Exploring The BUILDING BLOCKS of SCIENCE

Book 6
STUDENT TEXTBOOK

REBECCA W. KELLER, PhD

Illustrations: Janet Moneymaker
                 Marjie Bassler

Copyright © 2015 Gravitas Publications, Inc.

All rights reserved. No part of this publication may be reproduced, stored in a retrieval system, or transmitted, in any form or by any means, electronic, mechanical, photocopying, recording, or otherwise, without prior written permission from the publisher. No part of this book may be used or reproduced in any manner whatsoever without written permission.

Exploring the Building Blocks of Science Book 6 Student Textbook (softcover)

ISBN 978-1-941181-13-3

Published by Gravitas Publications, Inc.
Real Science-4-Kids®
www.realscience4kids.com
www.gravitaspublications.com

Contents

Introduction

CHAPTER 1 Technology in Science — 1
- 1.1 Introduction — 2
- 1.2 Archimedes: A Great Inventor — 3
- 1.3 How Science Shapes Technology — 5
- 1.4 How Technology Shapes Science — 6
- 1.5 Tools in Science — 8
- 1.6 Summary — 9

Chemistry

CHAPTER 2 Technology in Chemistry — 10
- 2.1 Introduction — 11
- 2.2 The Typical Chemistry Laboratory — 11
- 2.3 Types of Glassware and Plasticware — 13
- 2.4 Types of Balances and Scales — 16
- 2.5 Types of Instruments — 18
- 2.6 Summary — 20

CHAPTER 3 Acids, Bases, and pH — 21
- 3.1 Introduction — 22
- 3.2 Properties of Acids and Bases — 23
- 3.3 Acid-Base Theory — 23
- 3.4 Distinguishing Acids from Bases — 24
- 3.5 Acid-Base Indicators — 27
- 3.6 pH Meters — 28
- 3.7 Summary — 30

CHAPTER 4 Acid-Base Neutralization — 31
- 4.1 Introduction — 32
- 4.2 Titration — 33
- 4.3 Plotting Data — 34
- 4.4 Plot of an Acid-Base Titration — 36
- 4.5 Summary — 40

CHAPTER 5 Nutritional Chemistry 41

    5.1 Introduction 42
    5.2 Minerals 43
    5.3 Vitamins 45
    5.4 Carbohydrates 46
    5.5 Starches 47
    5.6 Cellulose 49
    5.7 Summary 50

Biology

CHAPTER 6 Technology in Biology 51

    6.1 Introduction 52
    6.2 The Botany Laboratory 53
    6.3 The Molecular Biology and Genetics Laboratory 54
    6.4 The Marine Biology Laboratory 56
    6.5 Summary 58

CHAPTER 7 The Microscope 59

    7.1 Introduction 60
    7.2 The Size of Things 61
    7.3 The Light Microscope 63
    7.4 The Electron Microscope 64
    7.5 Scanning Probe Microscopes 67
    7.6 Summary 70

CHAPTER 8 Protists 71

    8.1 Introduction 72
    8.2 Classification 73
    8.3 Photosynthetic Protists 76
    8.4 Heterotrophic Protists 78
    8.5 Summary 82

CHAPTER 9 Fungi: Molds, Yeasts, and Mushrooms 83

    9.1 Introduction 84
    9.2 Classification of Fungi 85
    9.3 Structure of Fungi 86
    9.4 Reproduction of Fungi 87
    9.5 Phylum Zygomycota—Molds 87
    9.6 Phylum Ascomycota—Yeasts, Truffles 89
    9.7 Phylum Basidiomycota—Mushrooms 92
    9.8 Summary 94

Physics

CHAPTER 10 Technology in Physics	95
10.1 Introduction	96
10.2 Some Basic Physics Tools	97
10.3 Mathematics	101
10.4 Electronics	103
10.5 Computers	106
10.6 CERN	107
10.7 Summary	109

CHAPTER 11 Motion	110
11.1 Introduction	111
11.2 Inertia	112
11.3 Mass	112
11.4 Friction	113
11.5 Momentum	114
11.6 Summary	116

CHAPTER 12 Linear Motion	117
12.1 Introduction	118
12.2 Speed	118
12.3 Velocity	124
12.4 Acceleration	128
12.5 A Note About Math	130
12.6 Summary	131

CHAPTER 13 Non-Linear (Curved) Motion	132
13.1 Introduction	133
13.2 Projectile Motion	133
13.3 Circular Motion	136
13.4 Summary	139

Geology

CHAPTER 14 Technology in Geology	140
14.1 Introduction	141
14.2 Hand Tools	141
14.3 Electronic Tools	143
14.4 Other Tools	145
14.5 Satellites	146
14.6 Summary	147

CHAPTER 15 Earth's Spheres   148
 15.1 Introduction   149
 15.2 The Spheres of Earth   150
 15.3 Connecting the Spheres   153
 15.4 A Delicate Balance   156
 15.5 Summary   158

CHAPTER 16 The Geosphere   159
 16.1 Introduction   160
 16.2 Minerals and Elements   162
 16.3 Using Volcanoes To See Inside Earth   163
 16.4 Using Earthquakes To See Inside Earth   166
 16.5 How Hot Is the Core?   170
 16.6 Summary   171

CHAPTER 17 The Atmosphere   172
 17.1 Introduction   173
 17.2 Chemical Composition   173
 17.3 Structure of the Atmosphere   175
 17.4 Atmospheric Pressure   178
 17.5 Gravity and the Atmosphere   180
 17.6 The Greenhouse Effect   180
 17.7 Summary   183

Astronomy

CHAPTER 18 Technology in Astronomy   184
 18.1 Introduction   185
 18.2 Telescopes   186
 18.3 Space Telescopes and Other Satellites   188
 18.4 Other Space Tools   190
 18.5 Summary   193

CHAPTER 19 Time, Clocks, and the Stars   194
 19.1 Introduction   195
 19.2 Reading a Star Atlas   197
 19.3 Time   199
 19.4 Celestial Clocks   202
 19.5 Summary   204

CHAPTER 20 Our Solar System   **205**

    20.1 Introduction   206
    20.2 Planetary Position   206
    20.3 Planetary Orbits   207
    20.4 Asteroids, Meteorites, and Comets   209
    20.5 Habitable Earth   212
    20.6 Summary   213

CHAPTER 21 Other Solar Systems   **214**

    21.1 Introduction   215
    21.2 Closest Stars   215
    21.3 Brightest and Largest Stars   217
    21.4 Planets Near Other Stars   218
    21.5 The Circumstellar Habitable Zone   220
    21.6 Summary   222

Conclusion

CHAPTER 22 Working Together   **223**

    22.1 Introduction   224
    22.2 Collaborating   226
    22.3 Global Collaboration   228
    22.4 Summary   232

Appendix: Math Solutions   233

Chapter 1 Technology in Science

1.1 Introduction 2

1.2 Archimedes: A Great Inventor 3

1.3 How Science Shapes Technology 5

1.4 How Technology Shapes Science 6

1.5 Tools in Science 8

1.6 Summary 9

Introduction

1.1 Introduction

Think for a moment about what it might be like to live in the 14th century. Imagine traveling back in time and finding yourself in a small European village in 1392.

What do you think you would find? How would you cook your food? Would you use an oven, a fire, or a microwave? How would you eat your food? Would you have a plastic cup for your milk? Would you have paper napkins and aluminum cans? How would you go from one city to the next? Could you get on a train? Would you have to walk or ride a horse?

How would you send a message to your mom to tell her that you'll be late for dinner? Could you email her or call her on your cell phone? How would you get clothes? Could you shop at a 14th century mall or on the internet? And what would your clothes be made of? Do you think you could find pink spandex shorts, or would they have to be made of brown cotton?

Many of the items you use today are the result of technology. Your cell phone, microwave oven, washing machine, and plastic cup all result from scientific discoveries combined with engineering. This combination of scientific discoveries and engineering has allowed people to invent products that have changed the way people live. Technological advances have improved our health, the food we eat, the clothes we wear, how we travel, and how we communicate with one another. Although technology has created some new problems (such as pollution), overall, technology has greatly improved many aspects of modern life.

The word technology comes from the Greek words *techne*, which means "craft," and *logy*, which means "the study of." So technology means the scientific study of craft. Craft, in this case, means any method or invention that allows humans to control their environment or adapt to it. Engineering is the branch of technology in which science and mathematics are used to design and build structures and machines.

1.2 Archimedes: A Great Inventor

How did technology get started? Although inventions and tool making have been around for as long as human beings have walked the Earth, technology started to develop when scientific thought (philosophy) and mathematics began to be used in combination with a scientific method of experimentation. This began sometime after the 15th century.

Archimedes of Syracuse (287-212 BCE) was an inventor in ancient Greece who combined engineering with science to develop different technologies. He is credited with inventing a machine called the Archimedean screw which is used for lifting water from one level to another. This device is made of a screw, or spiral, within a tube. As the screw turns, water within the spirals is lifted. The Archimedean screw is still used today in machines that move dry materials as well as those that move liquids.

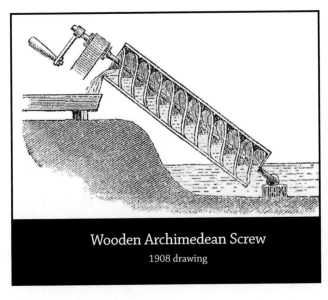

Wooden Archimedean Screw
1908 drawing

Archimedean screws being used at a German wastewater treatment plant

Archimedes is also credited with inventing the odometer, which measures the distance a vehicle has traveled. Archimedes made the odometer by first measuring the circumference of a vehicle's wheel and then calculating how many times the wheel would have to turn for the vehicle to travel a mile. By using gears attached to the wheel, a pebble was dropped into a box each time the wheel revolved the

number of times it would take to go a mile. At the end of the trip, the pebbles were counted to find the mileage.

The most famous story about Archimedes is the one in which he figured out how to determine whether the king's crown was made of solid gold. Archimedes had to figure out a way to test the crown in order to see whether it was made entirely of gold, and he had to do so without melting it. This was a puzzle for Archimedes and required a new way of thinking about the problem.

One day when Archimedes got into the bathtub, he noticed that his body displaced some of the bath water, making the water level rise. He realized he could use the displacement of water as a way to measure the crown's volume by measuring how much the water level changed when the crown was immersed. Once the volume of the crown was known, he could calculate the weight it should have been. By weighing the crown, he could then determine if it was solid gold or if it had some lighter weight metal, such as silver, mixed in with the gold. As the story goes, it was at this point that Archimedes jumped out of the tub and ran through the streets naked shouting, "Eureka! (I have found it!)" No one is certain whether the story is true, but it does give you an idea of how exciting new discoveries in science and technology can be!

1.3 How Science Shapes Technology

Before formal scientific disciplines (such as chemistry, biology, physics, geology, and astronomy) were defined, many early inventors tried to come up with ways to improve their lives by simply experimenting with items around them. Inventions and discoveries often happened by accident.

For example, although no one knows how glass making was discovered, it is thought that it was by accident. Perhaps when potters were firing their pottery, the ingredients for making glass were included in the objects they were making, and glass was discovered when the pots were heated to a high temperature. But no one really knows for certain. We do know that glass making was in existence more then 3400 years ago. Archeologists have found 3300 year old clay tablets with formulas and instructions for making glass. And an ancient shipwreck from 3400 years ago was discovered that contained glass ingots—chunks of glass that were meant to be remelted and formed into objects.

Regardless of how glass making was discovered, early glass makers could not have guessed that its discovery would help pave the way for Galileo and others to use telescopes for observing the stars.

The earliest star gazers had no way to see more than what they could observe in the sky with their unaided eyes. For these early astronomers, the technology for seeing the details of what is in and beyond our solar system did not yet exist. In the early 1600s, Galileo Galilei observed the heavens through the first telescope. However, in order for Galileo to use a telescope it had to be invented. The invention of the telescope required the development of the craft of glass making combined with the scientific discovery that glass could be formed into lenses to magnify distant objects.

Improvements in how much and how well a telescope could magnify were accelerated by Sir Isaac Newton's ideas. Newton thought about and tested his idea of using curved mirrors to focus light that enters a telescope and reflect the light to a lens, providing better quality images. Because Newton understood optics, the science of light, he was able to add to the technological advancement of the telescope.

SIR ISAAC NEWTON
1643-1727 CE

As we can see, science and technology work together. Without an understanding of basic scientific principles and without the gathering of new scientific facts, technological advances in any area would be impossible.

1.4 How Technology Shapes Science

In the last section we saw that science shapes technology, but how does technology shape science? Thinking back to the time of Galileo and Newton, how do you think the invention of the telescope has changed our understanding of not only our own solar system but also of the whole universe?

The invention of the telescope made amazing discoveries in astronomy possible. Before the telescope, no one had been able to observe details of the surface of the Moon and the planets, and no one knew how many planets were in

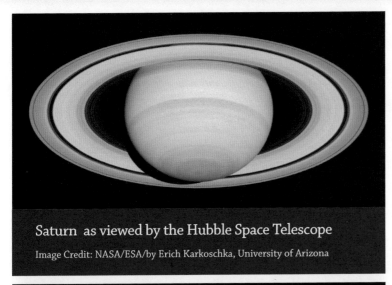

Saturn as viewed by the Hubble Space Telescope
Image Credit: NASA/ESA/by Erich Karkoschka, University of Arizona

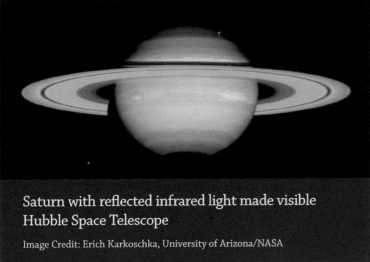

Saturn with reflected infrared light made visible
Hubble Space Telescope
Image Credit: Erich Karkoschka, University of Arizona/NASA

A view of the cosmos using space telescopes
Image Credit: NASA/CXC/SAO: X-ray;   NASA/JPL-Caltech: Infrared

our solar system. Without the technology of the telescope, much of what we have discovered about the universe would still be unknown.

The telescope has changed the way people think about cosmology, the study of the universe, and has opened up new areas of study in astronomy. It would be very difficult to study the universe if we had no way to look beyond our own planet. The technology of the telescope brought cosmology into the arena of serious scientific study. The telescope has enabled scientists to measure, predict, and quantify many features of the universe. Advances in technology have resulted in more powerful telescopes and telescopes traveling in space, giving us an increasingly detailed picture of the universe.

1.5 Tools in Science

Today, scientists in all scientific disciplines use tools and instruments that were developed as a result of technology shaping science and science shaping technology. For example:

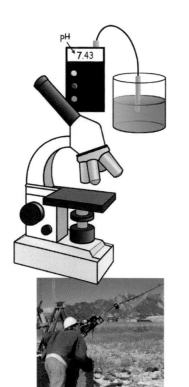

In chemistry, acids and bases are characterized by pH, and modern chemists often use digital pH meters to accurately measure pH. Chemists can follow changes in an acid-base reaction with a high degree of precision by using very sophisticated pH meters.

Biologists use a variety of different types of microscopes to see very small organisms. To see protists and cells, a biologist might use a light microscope, and to see even smaller structures, a biologist might use an electron microscope.

In physics, movement can be observed in great detail by using cameras, computers, and other equipment. Physicists can see precisely how the linear motion of a rolling ball changes when the ball encounters an obstacle, and the non-linear motion of a baseball or bullet can be measured.

Geologists use a variety of modern tools and instruments to study Earth. A field geologist is likely to study a mountain or a lava flow with a variety of hand tools and electronic instruments. Cameras, GPS devices, and small computers can be easily carried in a backpack alongside a rock hammer, compass, and hand lens.

Astronomers rely on complicated instruments to study the cosmos. Radio telescopes, space probes, rovers, and satellites all play a role in exploring the cosmos. It's even possible to put a probe on a fast-moving comet!

Photo credit: 1. Accelerometer, David Parsons/National Renewable Energy Laboratory
2. Mars rover, NASA/JPL/Cornell University

Science has driven the development of technological tools and instruments, and technology has expanded the development of scientific ideas and discoveries. Science and technology are interdependent with both being needed to advance our understanding of the world.

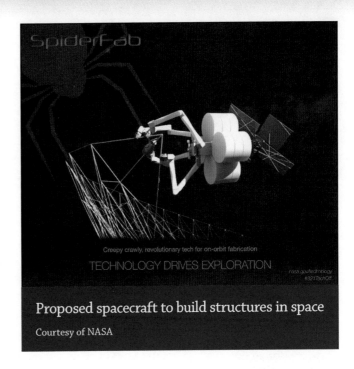

Proposed spacecraft to build structures in space
Courtesy of NASA

1.6 Summary

- The word technology means the scientific study of craft, or any method or invention that allows humans to control or adapt to their environment.

- Engineering is the branch of technology in which science and mathematics are used to design and build structures and machines.

- Archimedes is credited with being a great inventor, and the Archimedean screw is still in use today.

- Science has shaped technology and technology has shaped science.

- Science and technology are interdependent, and both are needed to study the world we live in.

Chapter 2 Technology in Chemistry

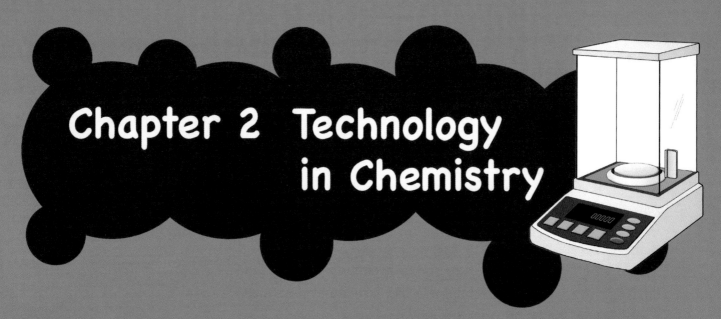

2.1 Introduction	11
2.2 The Typical Chemistry Laboratory	11
2.3 Types of Glassware and Plasticware	13
2.4 Types of Balances and Scales	16
2.5 Types of Instruments	18
2.6 Summary	20

2.1 Introduction

Chemistry has come a long way from the days of the alchemists and has changed dramatically within the last 100 years. Technological advances have changed how chemists study atoms, molecules, and chemical reactions. Today, chemists can image a single atom, use lasers to discover how molecules move, and work with sophisticated instruments to detect changes in chemical composition.

The tools used by modern chemists vary depending on the type of chemistry that is being studied. A

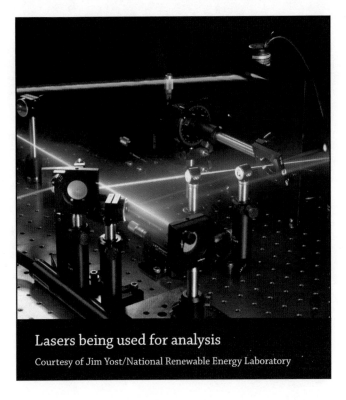

Lasers being used for analysis
Courtesy of Jim Yost/National Renewable Energy Laboratory

physical chemist may be interested in the motion of atoms and molecules and may have very different equipment than the organic chemist who is trying to create new compounds. The biochemist may use instruments similar to those found in a biology lab, and an analytical chemist may have a high-end analytical balance not found in a physical chemistry lab. However, whether the lab focuses on biochemistry, organic chemistry, or physical chemistry, most chemistry labs have some common basic equipment.

2.2 The Typical Chemistry Laboratory

Most chemistry labs contain certain basic equipment including different types of glassware for measuring and working with liquids, a balance or scale for measuring solids, a storage rack for chemicals, a stir plate for mixing liquids, bunsen burners with gas outlets for heating liquids and solids, a fume hood for containing sensitive or toxic materials, and safety equipment such as goggles, lab coats, and a safety shower where spilled chemicals can be washed off the body. Most chemistry labs also have long benches made of a material such as epoxy resin that can withstand heavy use and most chemical spills. Alongside the

12 Exploring the Building Blocks of Science: Book 6

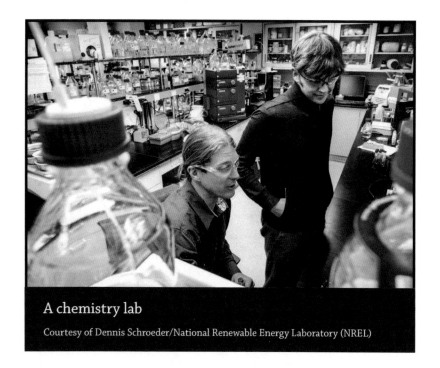

A chemistry lab
Courtesy of Dennis Schroeder/National Renewable Energy Laboratory (NREL)

benches are often small desks where research assistants work and keep their notebooks and important papers.

Some smaller equipment found in most chemistry labs includes various size spatulas and spoons for measuring chemicals, crucibles and crucible tongs for evaporating liquids, funnels, forceps, ring clamps, tube holders, and wash bottles.

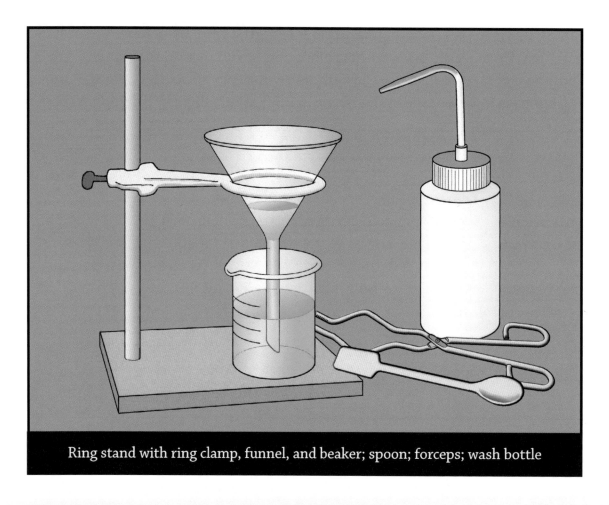

Ring stand with ring clamp, funnel, and beaker; spoon; forceps; wash bottle

2.3 Types of Glassware and Plasticware

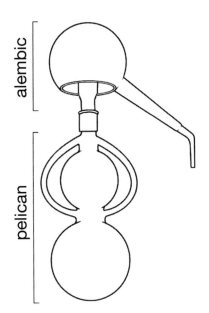

In the days of the alchemists, chemistry experiments were performed in specialized glassware. Two of the first types of glassware used by early chemists were the pelican and the alembic. The pelican and alembic were used for simple distillation, which is a process that separates a liquid mixture into its different individual components.

Distillation occurs when a mixture of liquids is heated and starts to vaporize, turning into a gas. As the vapors rise, they separate, with the lighter weight vapors moving through the long arm and the heavier vapors condensing and pouring back through the side arms.

Many labs now use a modernized version of the pelican and alembic. This version consists of a round bottomed flask that holds the liquid sample and is attached to a holder, a thermometer, and one end of a condenser. At the other end of the condenser is a collection flask. Since different liquids vaporize at different temperatures, this setup allows chemists to separate liquids by heating the liquid in the sample flask to the temperature at which one of the liquids in the mixture will vaporize. The vapor is returned to the liquid state in the condenser and flows into the collection flask.

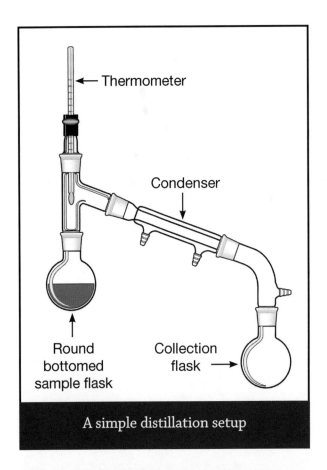

A simple distillation setup

Using beakers in an experiment
Courtesy of Pacific Northwest National Laboratory (PNNL)/DOE

Another type of modern glassware is the beaker. Beakers have a wide mouth and flat bottom and come in a variety of sizes. Beakers are used to measure and transfer large and small amounts of liquids. Beakers typically have markings along the side to indicate different volumes, and they have a spout for pouring.

Flasks are another type of glassware found in many chemistry labs. Flasks have a wide bottom and a narrow mouth. There are two main types of flat bottomed flasks. One type is the Erlenmeyer flask which is cone shaped with a broad, flat bottom and a narrow neck. Erlenmeyer flasks are named after the German chemist Emil Erlenmeyer. The narrow mouth of the Erlenmeyer flask helps minimize spills while mixing and can also be sealed with a rubber stopper.

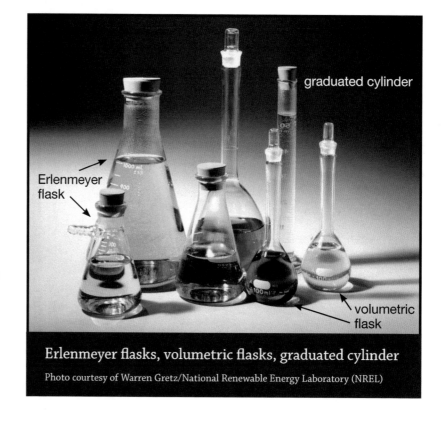

Erlenmeyer flasks, volumetric flasks, graduated cylinder
Photo courtesy of Warren Gretz/National Renewable Energy Laboratory (NREL)

Another type of flask is the volumetric flask which is round with a flat bottom and tall neck. Volumetric flasks are used to measure a single specific volume. On the neck there is a thin line or ring indicating when the flask is filled to an exact amount.

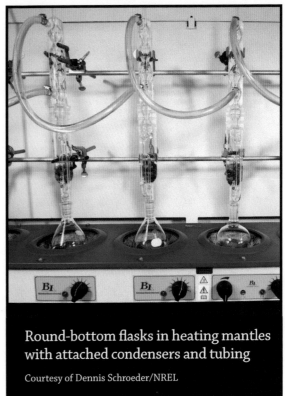

Round-bottom flasks in heating mantles with attached condensers and tubing

Courtesy of Dennis Schroeder/NREL

Round-bottomed flasks don't have a flat area on the bottom. These flasks are used for heating and collecting liquids in a distillation apparatus and can also be attached to other glassware. The rounded bottom helps heat to be equally distributed, especially if the flask is placed in a curved heating mantle.

Graduated cylinders are found in most chemistry labs and are used to accurately measure the volumes of liquids. A graduated cylinder is a long cylinder made of glass or plastic with markings drawn on the outside to indicate different volumes.

When small volumes or exact volumes need to be measured, a thin tube called a pipet can be used. There are different types of pipets for different purposes. The volumetric pipet is one type. A volumetric pipet is generally a long, thin tube with a bulb in the center. Volumetric pipets are used to measure a single specific volume and are very accurate.

A measuring pipet is a long, thin tube with markings along the side. Measuring pipets deliver various volumes with differing degrees of accuracy.

Transfer pipets are often disposable and generally made of plastic. They are used to transfer small amounts of liquid from one container to another and are not meant to be used for exact measurements.

Very small amounts of liquids can be measured using a pipetman with a plastic tip. A pipetman is a handheld piece of equipment that has a little pump inside that will measure very small amounts of liquids. When the pipetman is held in the palm of the hand and the pumping mechanism is depressed and then released, a small amount of liquid is drawn into the plastic tip. The tip can then be placed in a tube or other container and the liquid pushed out by the pumping mechanism.

Using a pipetman
Courtesy of US Department of Energy (DOE)

2.4 Types of Balances and Scales

Because chemical reactions require exact measurements of solid, powdered, and liquid materials, one of the most important tools in a chemistry lab is the balance or scale. Most labs use different types of balances or scales depending on the quantity to be measured and the accuracy needed.

The terms *balance* and *scale* are often used interchangeably in chemistry labs, but they are actually slightly different instruments. A balance measures the mass of an object by comparing the amount of force a sample exerts on a lever to the amount of force exerted by a standard reference, or known mass. A scale measures the weight of an object, with weight being related to the amount of the force of gravity that is pulling on the object.

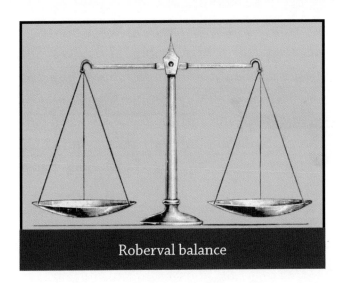
Roberval balance

Because balances measure mass directly, they are considered more accurate than scales. Many research and teaching laboratories have double-pan balances. This type of balance is called a Roberval balance because it was invented by the French mathematician Gilles Personne de Roberval (1602-1675 CE). It has two platforms, one on either side of a pivot point. When one side is heavier than the other, the pan on the heavier side will dip lower than the pan on the lighter side. This type of balance can be used to measure the mass of a sample by adding known reference weights to one pan until the reference weights balance with the sample on the other pan.

Although a balance is considered more accurate than a scale, scales can be used for most applications. Many modern chemistry labs have digital scales. Digital scales give measurements quickly and have an easy to read display. Digital scales can be battery powered or powered with an electric cord.

One type of digital scale used in many labs is the top-loading or pan scale. A pan scale is easy to load and can measure relatively large amounts of both dry and liquid materials. Typically, a measuring boat or beaker is placed on the scale and the scale is set to zero, or "zeroed," with the push of a button. This effectively subtracts the weight of the measuring boat or beaker from the total. Materials can then be added directly to the measuring boat or beaker and the digital pan scale will display the measurement.

If a more accurate measurement is needed and if the sample is small enough, an analytical balance can be used. An analytical balance has a pan mounted on an electromagnetic sensor, both of which are housed in a glass chamber. The glass chamber keeps small air currents from disturbing the measurement.

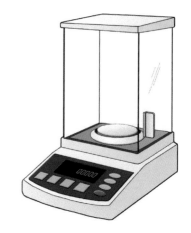

2.5 Types of Instruments

Depending on the type of research being done, many chemistry labs also have a number of specialized instruments that are used to analyze a variety of both chemical and physical properties.

For example, some labs have a gas chromatograph. A gas chromatograph is an instrument that can analyze mixtures of molecules that are volatile and can be turned into a gas. A gas chromatograph has a port where a small sample is injected with a syringe. The sample is heated quickly and turned into a gas. The gas travels down a column where the mixture is separated, then goes through a detector that identifies the gases in the sample. A recorder or computer creates an output showing the different compounds that are in the mixture.

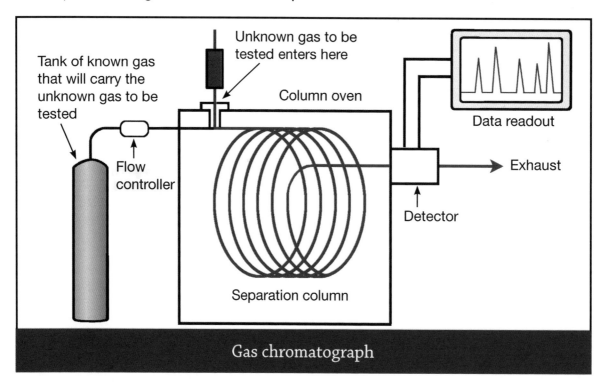

Gas chromatograph

Using a mass spectrometer
Courtesy of Dennis Schroeder/NREL

The detector in many gas chromatographs is another type of instrument called a mass spectrometer. A mass spectrometer is used to determine the mass of a given sample. In a combined gas chromatograph-mass spectrometer, once the gas has been separated, it enters the mass spectrometer where the gas molecules are bombarded with electrons. Different size atoms will give off different signals when hit with electrons, and these signals can be analyzed. A gas chromatograph-mass spectrometer can be used to test for drugs, toxins in water, and unknown samples.

Another type of instrument used to identify molecules is the infrared spectrometer. An infrared spectrometer measures the vibrations of atoms on a molecule. Molecules with lots of strong bonds and light atoms will vibrate differently than molecules with weak bonds and heavy atoms. This difference can be detected by the infrared spectrometer and can be used to identify different molecules and different types of bonds. Infrared spectrometers are common instruments in labs that study organic molecules.

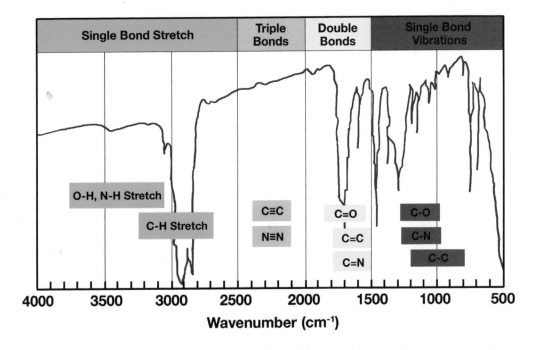

2.6 Summary

- Technology has helped equip modern chemistry labs with different types of glassware, balances, and other instruments.

- Flasks, beakers, and pipets are common types of glassware found in many chemistry labs.

- Double-pan balances, top-loading scales, and analytical scales are used in many chemistry labs.

- Although the type of chemistry research being done determines the instruments that will be needed, many labs have gas chromatographs, mass spectrometers, and infrared spectrometers.

Chapter 3 Acids, Bases, and pH

3.1 Introduction	22
3.2 Properties of Acids and Bases	23
3.3 Acid-Base Theory	23
3.4 Distinguishing Acids from Bases	24
3.5 Acid-Base Indicators	27
3.6 pH Meters	28
3.7 Summary	30

Chemistry

3.1 Introduction

Modern technology has allowed chemists to explore not only the structure, size, and properties of atoms and molecules but also how atoms and molecules interact with each other during chemical reactions. Two specific types of molecules that play a role in many important chemical reactions are acids and bases.

Even before people knew what acids and bases were, these compounds were used for many different purposes. Tablets from the Babylonian culture show that people knew how to make soap from bases 2800 years ago. Some of the Babylonian people and ancient German tribes used soap to style their hair. The base they used for soap making is called lye, which is a substance found in the ashes left over after wood has burned. Different kinds of oils or animal fats were then heated with the lye, producing soap.

ABU MUSA JABIR IBN HAYYAN
Circa 721-815 CE

Some acids were also known many centuries ago. An Iranian alchemist named Abu Musa Jabir ibn Hayyan (circa 721-815 CE) discovered hydrochloric acid by mixing a salt (sodium chloride) with sulfuric acid. Jabir ibn Hayyan also developed aqua regia (royal water) by mixing nitric acid and hydrochloric acid. This material could easily dissolve gold and was often used to determine whether or not a substance appearing to be gold was, in fact, gold.

Chemistry—Chapter 3: Acids, Bases, and pH

3.2 Properties of Acids and Bases

Some general properties of acids and bases are listed in the following chart. Vinegar, tomatoes, and black coffee are all acids, and all have a sour taste. It turns out that most acids are sour tasting. Grapefruit, for example, can be very sour. The juice inside a grapefruit contains a lot of citric acid.

General Properties	
Acids	Bases
• Sour in taste • Not slippery to the touch • Dissolve metals	• Bitter in taste • Slippery to the touch • React with metals to form precipitates

Detergents and many cleaners feel slippery. This is because many cleaners are basic and slipperiness tends to be a property of bases.

Some acids and bases are very poisonous or corrosive and can easily harm you. Battery acid, for example, can burn your skin if it happens to spill and can also make you very sick if eaten.

3.3 Acid-Base Theory

Because acids and bases are so important, chemists have developed several ways of understanding them. In 1834, Michael Faraday (1791-1867 CE) discovered that acids and bases are electrolytes, meaning they form ions when dissolved in water and can conduct electricity. Ions are atoms or molecules that have an electric charge.

Svante Arrhenius (1859-1927 CE), a Swedish chemist, took the next step in understanding acids and bases. In 1884 Arrhenius showed that acids produce hydrogen ions (H+) in water and bases produce hydroxide ions (OH-) in water. This is a useful theory for

SVANTE ARRHENIUS
1859-1927 CE

explaining acids and bases, and the Arrhenius definitions are still widely used today. By definition, an Arrhenius acid is any molecule that releases a hydrogen ion (H+), and an Arrhenius base is any molecule that releases a hydroxide ion (OH-). When using this definition, keep in mind that it only applies to hydrogen and hydroxide ions in aqueous (water) solutions. Acids and bases can also be

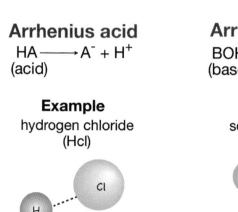

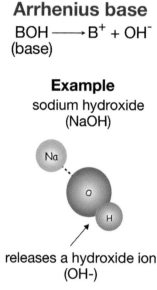

defined as the giving or taking of protons or electrons in non-aqueous (non-water) solutions. However, for the chemical reactions explored in this textbook, the Arrhenius definition of an acid and base will be used.

3.4 Distinguishing Acids from Bases

The use of litmus paper was the first method discovered for determining whether a liquid is an acid or a base. The word litmus comes from an old Norse word meaning "to dye or color." Certain species of lichens provide the dye used in litmus paper. A lichen is an organism that consists of a fungus and algae working in partnership to form the organism. Litmus paper was first used by the Spanish alchemist Arnaldus de Villa Nova (circa 1235-1311 CE) to test whether a substance was an acid or a base.

Blue litmus paper reacts with an acid. Red litmus paper reacts with a base.

Blue litmus paper will turn red in the presence of an acid and red litmus paper will turn blue in the presence of a base. Because litmus paper is relatively inexpensive to produce and easy to use, it can be used in the chemistry lab to quickly determine whether an aqueous solution is an acid or a base. Litmus paper is a great tool for chemical geologists when they are out in the field and need an inexpensive and lightweight method for testing the water in ponds, rivers, and geothermal pools.

Litmus paper can be perfect for determining whether a solution is an acid or a base, but it cannot measure how concentrated an acid or a base is. Concentration is defined as the number of units in a given volume. The definition of a unit can vary, and the unit can be an atom, an electron, or even a ping-pong ball. A solution that contains many units is called concentrated (or strong), and a solution with few units is called dilute (or weak). For example, a concentrated solution of hydrochloric acid (HCl) has many HCl molecules, and a dilute solution of HCl has few HCl molecules.

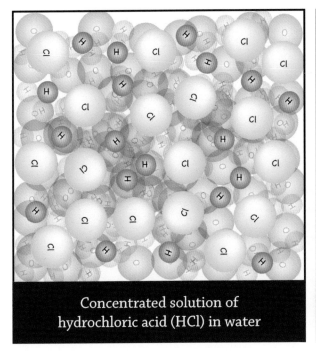

Concentrated solution of hydrochloric acid (HCl) in water

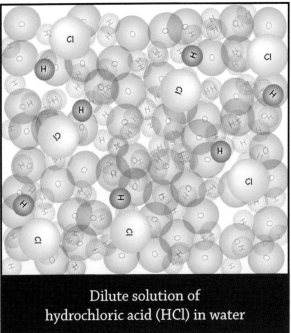

Dilute solution of hydrochloric acid (HCl) in water

The concentration of an acid or base depends on the number of acid or base ions in a solution. Using the Arrhenius definition, a concentrated acid is an acid that contains a large number of hydrogen ions (H+) and a concentrated base is a base with a large number of hydroxide ions (OH-). Conversely, a dilute acid is an acid with few hydrogen ions and a dilute base is a base with few hydroxide ions. But how do you measure the number of hydrogen or hydroxide ions in a given volume?

In 1909 while working for a brewery in Sweden, Sören Peter Lauritz Sörensen (1868-1939 CE), a Danish chemist, introduced the pH scale (pH is pronounced "P" "H"). The pH scale makes it easier for chemists to describe how many hydrogen ions are in a solution, and therefore, how acidic or basic a solution is. For the brewery, knowing the pH of their mixtures enabled them to control the acidity and make a better and more consistent product.

Specifically, pH is a measure of the concentration of hydrogen ions in a solution. According to the pH scale, an acid has a pH below 7 and a base has a pH above 7. Neutral water has a pH equal to 7.

> pH = 7: The solution is neither an acid nor a base—it is neutral.
> pH less than 7: The solution is an acid.
> pH more than 7: The solution is a base.

The following chart shows the pH for different solutions.

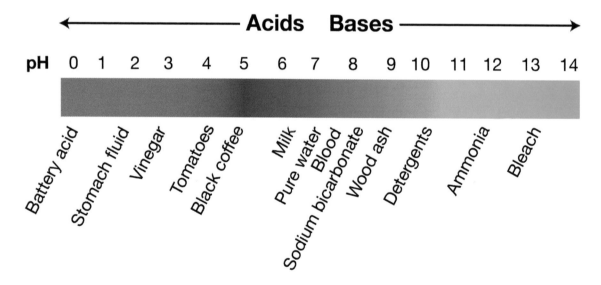

Notice that the pH for blood is near 7, close to the pH of water. Our bodies are made mostly of water, and blood carries nutrients throughout our bodies. It is important that the pH of blood be near 7 since many of our cells and tissues would be damaged if the pH were much higher or much lower than 7. However, notice that the pH for stomach fluid is even lower than the pH for vinegar. Why is stomach fluid so acidic? As it turns out, your stomach makes hydrochloric acid (HCl). This acid helps break down your food so that it can be carried to other places in your body. The inside of your stomach has a special lining that is designed to prevent the acidic stomach fluid from causing damage.

3.5 Acid-Base Indicators

Sörensen was able to create a pH scale by using a set of indicators that change color as the pH changes. Litmus paper is one type of acid-base indicator.

There are also other kinds of acid-base indicators. Some indicators change colors at very low pH, and others don't change until the pH is very high. Some indicators even change colors twice. The following chart shows a few acid-base indicators and the pH range in which they change color.

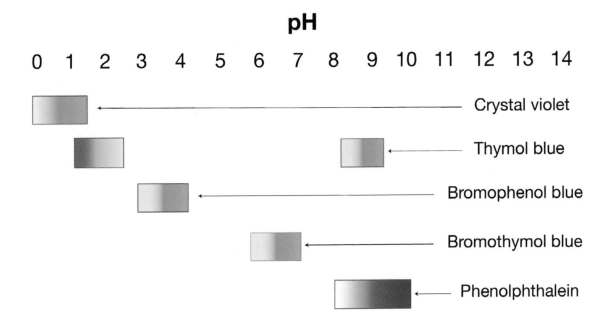

Notice that crystal violet changes colors at very low pH. Below pH 1, crystal violet is yellow, but above pH 1 it is blue. Now look at phenolphthalein. Below pH 9 phenolphthalein is not colored, but above pH 9 it turns pink. This narrow range of some pH indicators is very useful for finding the pH of a solution and also determining how concentrated the acid or base is.

28 Exploring the Building Blocks of Science: Book 6

In general, any molecule that changes color when the pH changes can be considered an acid-base indicator. There are many different acid-base indicators, including red cabbage juice which is an acid-base indicator that is easy to make and fun to use.

3.6 pH Meters

Litmus paper and other indicators make it relatively easy to tell whether a solution is an acid or a base, but measuring hydrogen ion concentration and determining the strength or weakness of an acid or base can be more complicated.

Recall from Section 3.3 that Michael Faraday discovered that acids and bases are electrolytes that conduct electricity. A pH meter works by measuring electrical resistance in a water solution (how well electricity moves through the solution). A pH meter has a probe, or electrode, that is usually made of glass and is connected to a meter that measures how much electricity an acid or base can conduct.

In 1906 two scientists, Fritz Haber (1868-1934 CE) and Zygmunt Klemensiewicz (1886-1963 CE) tried to create the first pH meter. They made glass probes and attempted to measure pH directly by inserting a probe into a solution, but the glass needed to be very thin and the probes broke easily. They were never able to get their early pH meter to work.

ARNOLD BECKMAN
1900-2004 CE

In 1934 Arnold Beckman (1900-2004 CE) became the inventor of the first successful pH meter. Beckman was a chemistry professor at the California Institute of Technology (Caltech) when he was asked by the California Fruit Growers Exchange to find a way to measure the acidity of lemon juice. Members of the California Fruit Growers Exchange grew most of the citrus fruit in California at that time, and they needed a quick and easy way to see how acidic the fruit was. This information helped them decide when the fruit was ready to harvest.

After Beckman had perfected the first pH meter, he went into the business of producing and selling them. The first pH meters went on the market in 1935. Many people believed that only about 600 pH meters would be needed to supply chemistry labs around the world, but Beckman proved them wrong. Over the next couple of decades, Beckman's company grew, developing and selling other scientific instruments as well as the pH meter, and he eventually became a millionaire. Beckman was very generous with his money, contributing over 400 million dollars to science research and education during his lifetime.

Chemists now use both portable and stationary pH meters. A portable pH meter can be transported easily in a backpack or pocket and can be used outdoors to measure the pH of rivers, ponds, geothermal pools, and other water sources. Stationary pH meters, also called benchtop meters are usually more accurate than portable pH meters and are used inside a laboratory.

3.7 Summary

- Acids are generally sour in taste, not slippery to the touch, and dissolve metals.

- Bases are generally bitter in taste, slippery to the touch, and form precipitates with metals.

- Concentration is defined as the number of units in a given volume.

- Indicators such as litmus paper can be used to determine if a solution is an acid or a base.

- The pH of a solution measures the ion concentration.

- pH can be measured by pH meters, pH paper, and acid-base indicators.

Chapter 4 Acid-Base Neutralization

4.1 Introduction	32
4.2 Titration	33
4.3 Plotting Data	34
4.4 Plot of an Acid-Base Titration	36
4.5 Summary	40

Chemistry

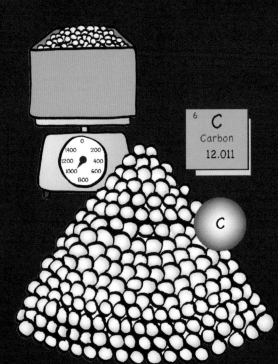
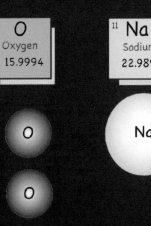
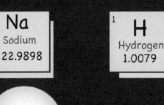

4.1 Introduction

When an acid is added to a base, or a base is added to an acid, an acid-base reaction occurs. An acid-base reaction is a special type of exchange reaction. Recall that in an exchange reaction atoms in one molecule trade places with atoms in another molecule.

An example of an acid-base reaction is the chemical reaction between vinegar and baking soda. Vinegar (acetic acid) is an acid, and baking soda (sodium hydroxide) is a base. When vinegar and baking soda are mixed in water, one of the hydrogen atoms of the vinegar trades places with the sodium atom of the baking soda. However, there is a second step—a decomposition reaction in which the intermediate products break apart and make carbon dioxide and water. You can see that the end product of the reaction between vinegar and baking soda is a solution of water and sodium acetate (a salt), with carbon dioxide being given off as gas bubbles.

Acid-base exchange reaction

acetic acid (vinegar) + sodium bicarbonate (baking soda)

A hydrogen ion from the vinegar breaks away and replaces the sodium ion on the baking soda.

The vinegar picks up the free sodium ion and forms sodium acetate.

sodium acetate

Decomposition reaction

hydrogen ion

hydroxyl ion

water

Carbon dioxide is released as a gas and bubbles away.

The sodium acetate stays in the solution dissolved in the water.

An acid-base reaction is also called a neutralization reaction. When an equal concentration of a strong acid reacts with an equal concentration of a strong base, the resulting solution becomes neutral—neither acidic nor basic—because all the molecules of both the acid and the base have reacted with each other.

Another way of saying this is that when an acid reacts with a base, the atoms that make the acid acidic (hydrogen ions) react with the atoms that make the base basic (hydroxide ions) to form water and a salt. When this happens, the acid and base neutralize each other.

As an example of an acid being neutralized, you might think about a person who gets acid indigestion after eating too many chili cheese fries. Acid indigestion results when the stomach has produced too much acid and the excess acid causes pain. Taking an antacid can reduce the pain because an antacid is a base and when eaten will neutralize stomach acid. Antacids are not very strong (not concentrated) so they are safe to eat when the need arises.

4.2 Titration

Why do we need to take antacids that are not very concentrated? How can we be sure that the antacid we take is not too strong but just strong enough to fix our indigestion? If we have an acid or a base, how can we know whether it has a concentration that is safe for us to use?

As it turns out, chemists can determine how concentrated and how strong an acid or a base is by using a technique called a titration. The word titration comes from the old French word *titre* which means assay. An assay is a test of quality. By doing an acid-base titration, a chemist can test for the quality of an acid or base solution—how strong, weak, or concentrated it is.

The fundamental idea behind a titration is to use a solution of an acid or base of known concentration to find the concentration of an unknown acid or base solution. An acid-base indicator or a pH meter is used to observe the acid-base reaction during the titration.

To do a titration, an acid is added to a base or a base is added to an acid, and the pH of the solution is observed as it changes. If the acid or base is added a little at a time, the pH of the solution will change slowly enough to be observed by a pH meter or acid-base indicator.

For example, if the titration starts with a beaker full of vinegar and red cabbage juice acid-base indicator, the color of the liquid will be a bright pink. If one spoonful of baking soda is added, there is not enough base to completely neutralize the acid, so the solution will still be pink. However, if more spoonfuls of baking soda are added one at a time, eventually there will be enough base to neutralize the acid, and the acid-base indicator will change color.

By plotting the data on a graph, the concentration of the unknown acid or base can be determined. Before we see how this works, let's take a look at how to plot data.

An acid; no base added.

One spoonful of base added, but the solution is still acidic.

The solution starts to change color as more spoonfuls of base are added.

Eventually the solution changes color completely as the acid is neutralized.

4.3 Plotting Data

One way to examine a titration in detail is to plot or graph the data. A plot is a handy visual tool that scientists use to help them understand data. Plots can be made of almost any data. For example, you might notice that the older members in your family are usually taller than the young members, so you could say that there is a connection between age and height. A plot can be made to illustrate the relationship between age and height.

To make a plot of age vs. (versus) height, the first step is to gather the data to be plotted. For this example, the data might look something like the following.

Age	Height
Age 1	.6 m (2 ft)
Age 6	.9 m (3 ft)
Age 8	1.2 m (4 ft)
Age 11	1.5 m (5 ft)
Age 30	1.7 m (5.5 ft)
Age 40	1.8 m (5.8 ft)
Age 60	1.75 m (5.75 ft)

Once the data has been collected, the plot can be created. First, a horizontal line is drawn. This line is called the x-axis. Another line is drawn perpendicular to the first (vertically) and meets the first line at the bottom of the left-hand side. This vertical line is called the y-axis.

To plot the data in this example, the x-axis is labeled "age" and the y-axis is labeled "height." The age of the person is marked on the graph with a vertical dotted line, and the height of the person is marked with a horizontal dotted line. The point where the two lines intersect is marked with a red dot, called a point. A solid black line can then be drawn to connect all the points on the plot. From this plot we can tell that, in general, as a person gets older, they grow taller. (The drawn black line goes up as the age goes up.) This graph also shows that a person stops growing when a certain age is reached. (The drawn black line levels off, showing that no significant growth occurs after age 20 or so.) Plotting is a tool that scientists use to organize their data in a way that makes it easier to understand.

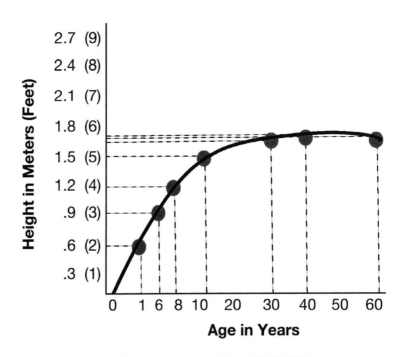

4.4 Plot of an Acid-Base Titration

Let's take a closer look at how to plot an acid-base titration and how to find out the concentration of an unknown acid or base. Imagine that we have a beaker half-full of household vinegar. We know that household vinegar is made of acetic acid molecules, but we may not know how many acetic acid molecules the beaker contains. In other words, we don't know how concentrated our household vinegar is. Imagine that we also have a box of baking soda. Using a scale we can measure the weight of the baking soda. By knowing the weight of the baking soda, we know how many molecules it has.

Wait! How can we know this? Let's use a box of ping-pong balls as an example. Imagine you have a box full of ping-pong balls. You don't know how many ping-pong balls you have, but you can find out the weight of one ping-pong ball. You can then calculate the total number of ping-pong balls. First, by dumping

out all the ping-pong balls and weighing the empty box, you can get the weight of the box alone. Next, if you put all the ping-pong balls back in the box and weigh the full box, you can get the total weight. Now you can subtract the weight of the empty box from the total weight of the box with the ping-pong balls in it to find out the weight of all your ping-pong balls together.

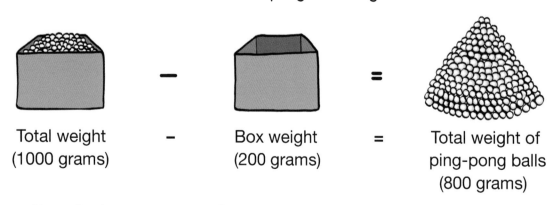

Total weight — Box weight = Total weight of
(1000 grams)   (200 grams)   ping-pong balls
                             (800 grams)

If you then divide the weight of all the ping-pong balls by the weight of one ping-pong ball, you can calculate how many ping-pong balls you have.

Chemistry—Chapter 4: Acid-Base Neutralization

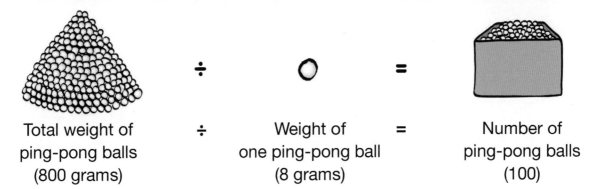

| Total weight of ping-pong balls (800 grams) | ÷ | Weight of one ping-pong ball (8 grams) | = | Number of ping-pong balls (100) |

The same is true for calculating how many molecules you have in a given amount. If you have a teaspoonful of baking soda that weighs 100 grams and you know the weight of one baking soda molecule, you can figure out the number of baking soda molecules in 100 grams.

But wait! How do we know the weight of one baking soda molecule? We can't weigh a baking soda molecule directly, but we do know that one baking soda molecule contains one carbon atom, three oxygen atoms, one sodium atom, and one hydrogen atom. Because we can use the periodic table of elements to find the atomic weight of each atom, we can add the atomic weights of all the atoms together to get the molecular weight—the weight of one molecule.

We can see that the molecular weight of sodium bicarbonate is 84 amu. But wait! What is an amu and how many grams is that? An amu is a measure of the atomic mass of an atom. Amu stands for atomic mass units. An atomic mass unit is equal to 1/12th the mass of a carbon atom.

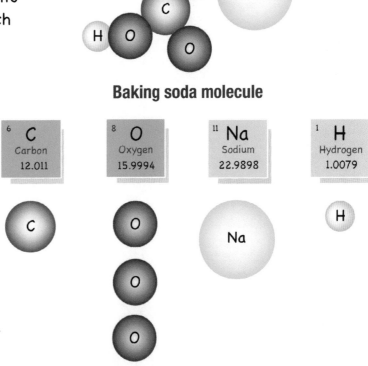

Baking soda molecule

| ^{6} C Carbon 12.011 | ^{8} O Oxygen 15.9994 | ^{11} Na Sodium 22.9898 | ^{1} H Hydrogen 1.0079 |

1 x 12 amu + (3 x 16 amu) + (1 x 23 amu) + (1 x 1 amu) = (84 amu)

As you can imagine, one atomic mass unit is a very small number. In fact, this number is so small that it is not useful for chemists doing chemistry calculations. Instead of using atomic mass units, chemists figured out how to measure the weight of a group of atoms. This group of atoms is called a mole. A mole is just a name that is used to represent a certain number of "things." These "things" can be atoms, molecules, ions, or even oven mitts or baseball caps. A mole is a way to count atoms and molecules just like a dozen is a way to count eggs. The only difference is that a mole is a very large number.

$$\text{One mole} = 602{,}200{,}000{,}000{,}000{,}000{,}000{,}000$$
$$(6.022 \times 10^{23})$$

A mole is 6022 followed by 20 zeros, and it's such a big number that it won't even fit on most calculators. It's so big that if you had a mole of marbles, it would be bigger than the Moon! But atoms are very tiny, so a mole of atoms is a nice, manageable size. A mole of most atoms will fit in the palm of your hand.

AMEDEO AVOGADRO
1776-1856 CE

This big number for one mole is called Avogadro's constant and is named after Italian scientist Amedeo Avogadro who had the idea that the number of gas molecules in a given volume is the same no matter what kind of gas it is. Based on his idea, Avogadro was able to calculate the number of molecules in the given volume. Today, Avogadro's number is used to relate the number of atoms and molecules to their atomic and molecular weights.

By definition, one mole of carbon atoms equals 12 grams, one mole of hydrogen atoms equals 1 gram, and 1 mole of sodium bicarbonate molecules equals 84 grams. This comes in very handy for acid-base reactions. Instead of worrying about the number of molecules needed to neutralize a reaction, we can use the number of moles. One mole of sodium bicarbonate will neutralize one mole of vinegar. We know that one mole of baking soda weighs 84 grams and we can measure this on a scale!

Now that we know that 84 grams of baking soda equals one mole, we can use a titration to find out how many moles of acid are in a solution. Let's go back to the titration of vinegar and baking soda. We have a beaker half-full of vinegar and we have a box of baking soda. Using a scale we can measure how much a teaspoonful of baking soda weighs in grams. We also know that one mole of baking soda will neutralize one mole of vinegar.

By using a pH meter or an acid-base indicator, we can see the pH of the solution change as we slowly add baking soda to the vinegar. We can record the pH change for each teaspoon of baking soda added and then plot a graph from this data. The graph may look something like the one following.

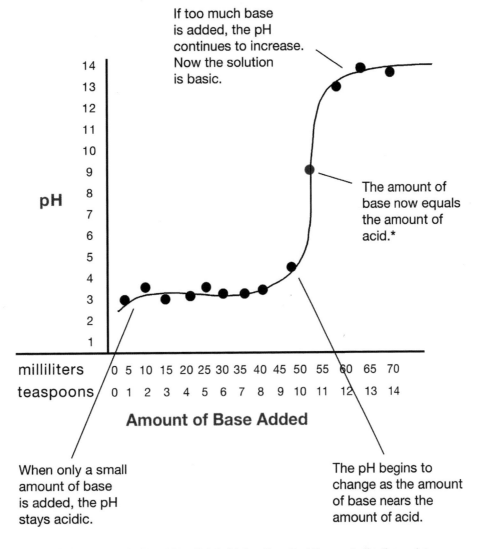

* In this example the pH is slightly higher than 7 at the neutralization point because the product of the reaction, sodium acetate, is slightly basic.

The x-axis (horizontal) is labeled *Amount of Base Added*, and the y-axis (vertical) is labeled *pH* but could also be labeled *Color*. Notice that the graph looks like a snake with two curved sections. The midpoint between the two curved sections is the point where the amount of base equals the amount of acid. If you know how many moles of base are in a teaspoonful and you know how many teaspoonfuls were added, you know how many moles of base it takes to neutralize *all* the acid. Therefore, you know how many moles of acid were in the beaker you started with!

> ## You Do it!
>
> 1. If it takes 84 grams of baking soda to neutralize a beaker of acetic acid, how many moles of acetic acid do you have?
>
> 2. If if takes 42 grams of baking soda to neutralize a beaker of acetic acid, how many moles of acetic acid do you have?
>
> 3. If if takes 168 grams of baking soda to neutralize a beaker of acetic acid, how many moles of acetic acid do you have?
>
> (See Appendix for solutions.)

4.5 Summary

- An acid-base reaction is a special type of exchange reaction.
- An acid and a base neutralize each other in an acid-base reaction.
- Equal amounts of acid and base completely neutralize each other.
- Plotting data can make it easier to understand.
- A mole is the name for a certain number of atoms, molecules, or ions.
- The concentration of an unknown acid or base can be found using a titration.

Chapter 5 Nutritional Chemistry

5.1	Introduction	42
5.2	Minerals	43
5.3	Vitamins	45
5.4	Carbohydrates	46
5.5	Starches	47
5.6	Cellulose	49
5.7	Summary	50

5.1 Introduction

We saw in the last chapter how acid-base reactions can help neutralize stomach acid, but what other reactions occur in the body? Where do we get the molecules we need to live? What happens to the food we eat? We may know from experience that when we skip a meal or are unable to eat because of illness, we become weak and lack energy. Our bodies require food to help us grow and keep our "engines running." Without food we would not survive.

Unlike plants, we cannot stick our feet in the soil, lift our hands to the Sun, and make our own food. In fact, we rely on plants and animals to provide the food our bodies need to keep us going. But what is in food and what does food do for us?

There are many different kinds of food the body requires to stay healthy. These foods provide important nutrients including vitamins, proteins, fats, and carbohydrates. We get these important nutrients from eating a variety of foods. The nutrients we get from eating foods provide the necessary molecules our bodies need to grow and function properly. Vitamins, like those found in carrots, help our eyes work. Fats found in vegetable oil and butter help our brain and other tissues function. Proteins from fish and meats help our bones heal and our muscles grow. Carbohydrates, like those found in bread, potatoes, and sweets, provide us with energy.

5.2 Minerals

The smallest nutrients found in many of the foods we eat are minerals. Minerals are salts of various elements. Minerals are not manufactured inside living things but are found in the Earth's soil. Plants get their minerals directly from the soil, and animals get most of their minerals from plants or other animals. Humans require a modest amount of seven different elements found in minerals (calcium, phosphorus, potassium, chlorine, sodium, magnesium, and sulfur) and extremely small, or trace, amounts of several other elements, such as fluorine, cobalt, copper, and iron.

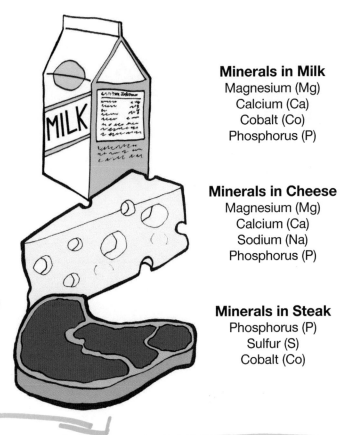

Minerals in Milk
Magnesium (Mg)
Calcium (Ca)
Cobalt (Co)
Phosphorus (P)

Minerals in Cheese
Magnesium (Mg)
Calcium (Ca)
Sodium (Na)
Phosphorus (P)

Minerals in Steak
Phosphorus (P)
Sulfur (S)
Cobalt (Co)

Elements such as calcium (Ca), magnesium (Mg), phosphorus (P), sulfur (S), and cobalt (Co) are found in foods such as cheese, milk, and meat.

Minerals are not used for fuel but are used to help maintain many of the systems inside our cells. For example, calcium, magnesium, and phosphorus help strengthen bones and harden teeth; iron is essential for oxygen binding in blood cells; and copper is important for the correct functioning of nerve tissues.

We get the different minerals we need from a variety of different foods. Calcium and phosphorus are found in milk products, leafy vegetables, and egg yolks. Sodium, potassium, and chlorine are found in table salts, cheese, and dried apricots. Magnesium is found in whole-grain cereals and a variety of leafy green vegetables. Iron is found in meat, dried nuts, and molasses; and zinc is found in seafood, nuts, and yeast. Without minerals our bodies would not function properly. The lack of necessary minerals can result in a variety of diseases or even death.

Below is a table showing some of the minerals our bodies need, what foods they are found in, and how the body uses them.

Mineral	Sources	Uses in Body
Potassium (K)	avocados, apricots, meats, leafy greens	muscle contraction, making proteins
Sodium (Na)	table salt, cheese, cured meats, celery	maintains water balance and cellular pumps
Calcium (Ca)	milk, milk products, egg yolk, leafy greens	needed for bones and teeth, blood clotting, nerve impulses
Chlorine (Cl)	table salt, seaweed, tomatoes, lettuce	activates salivary amylase, helps transport CO_2
Magnesium (Mg)	milk, dairy products, whole grain cereals, nuts, spinach	required for normal muscle and nerve function, bones
Sulfur (S)	meat, milk, eggs, legumes, garlic, onions	found in some proteins, needed for cartilage, bones, and tendons
Phosphorus (P)	milk, eggs, meat, fish, nuts, whole grains	needed for bones and teeth, nerve activity, and energy storage
Cobalt (Co)	liver, lean meat, fish, milk, broccoli	needed for vitamin B_{12}
Fluorine (Fl)	fluoridated water	tooth structure, prevents dental cavities, and may prevent osteoporosis
Copper (Cu)	liver, shellfish, whole grains, meat, beans, nuts, potatoes	needed for the manufacture of nerve tissues and blood molecules
Iodine (I)	iodized salt, shellfish, cod liver oil, milk, eggs, seaweed, potatoes	needed for making thyroid molecules
Manganese (Mn)	nuts, whole grains, fruit, leafy vegetables	needed for making fats, blood molecules, and carbohydrates
Iron (Fe)	meat, liver, nuts, egg yolk, molasses, beans, spinach	essential part of blood molecules that bind oxygen
Selenium (Se)	seafood, meat, cereal, milk	component of certain proteins
Zinc (Zn)	seafood, nuts, yeast, cereal, meat, beans	component of some proteins, required for wound healing, taste, and smell
Chromium (Cr)	liver, meat, broccoli, yeast, fruit, whole grains	component of some proteins and needed for glucose use

5.3 Vitamins

Vitamins are another set of molecules needed for a healthy body. Humans require a variety of vitamins for growth and good health. Like minerals, vitamins are not used as sources of energy but instead function as "helper" molecules for several different chemical reactions that occur in the body. Like minerals, most vitamins are found in the food we eat. However, human bodies do make two vitamins—vitamin D which is made in our skin and vitamin K which is made by beneficial *E. coli* bacteria that live in our intestines. Because there is no one food that contains all of the vitamins necessary for the healthy maintenance of the body, it is important to eat a variety of foods.

Below is a table showing some of the vitamins we need, what foods they are in, and how the body uses them.

Vitamin	Sources	Uses in Body
Fat-soluble		
Vitamin A	yellow and green vegetables, fish liver oil	needed for normal tooth and bone development, eyes, skin
Vitamin E	wheat germ, vegetable oil, dark green vegetables, nuts	protects cell membranes and prevents hardening of arteries
Vitamin D	produced in the skin by ultraviolet light, also in cod liver oil, egg yolk, milk	needed for normal tooth and bone development, blood clotting
Vitamin K	made by bacteria inside the body, also in cabbage, liver, green leafy vegetables	needed to form some proteins and for cell function
Water-soluble		
Vitamin C	fruits, vegetables, tomatoes, potatoes	used in the formation of all connective tissue, helps iron absorption
Vitamin B_1	lentils, green peas, pork, whole grains	needed for making certain sugars
Vitamin B_2	meats, milk, egg white, green leafy vegetables, legumes	needed for some protein function, making red blood cells
Vitamin B_{12}	liver, meat, fish, dairy foods, eggs	needed for the nervous system, bone marrow, and making DNA
Vitamin B_6	liver, fish, bananas, sweet potatoes	needed for making DNA and certain proteins
Vitamin B_5	liver, eggs, meat, legumes, potatoes, milk	needed to make fats, steroids, and blood molecules

Some of the vitamins we need are soluble (will dissolve) in water, and the body absorbs these vitamins directly when the digestive tract absorbs water. However, some vitamins are soluble only in fat, and the body absorbs these vitamins when we eat fats. For example, vitamin A is an important vitamin that is only soluble in fats, so it is important to eat enough fats for the body to get the vitamin A it needs. Fat-soluble vitamins are stored by the body but water-soluble vitamins are not. Our body eliminates excess water-soluble vitamins, but because our bodies store fat-soluble vitamins, it is possible to get too much of these. For example, too much vitamin A is toxic and can cause nausea, vomiting, or bone and joint pain.

5.4 Carbohydrates

Carbohydrates are another set of molecules essential for living things. Carbohydrates are the most abundant class of biological molecules and are found in every living thing. The word carbohydrate comes from the name of the element carbon and the Greek word *hydor*, meaning "water," so a carbohydrate is a molecule made of both carbon and water.

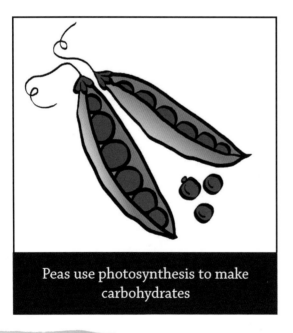

Peas use photosynthesis to make carbohydrates

Carbohydrates are made inside living things through two main biochemical processes: gluconeogenesis and photosynthesis. Gluconeogenesis comes from the Greek words *glykys* which means "sweet," *neo* which means "new," and *gen* which means "birth" or "produce." Gluconeogenesis literally means the "new production of sweet molecules." Gluconeogenesis occurs in the liver and kidneys of humans, and it is the biochemical pathway that makes carbohydrates when no food is consumed (for example, during a fast). Photosynthesis, on the other hand, is the biochemical process that plants use to convert light energy into food energy, or sugars. The bulk of carbohydrate molecules come from photosynthesis.

Chemistry—Chapter 5: Nutritional Chemistry

The simplest carbohydrates are the sugars. Sugars are relatively small molecules. They taste sweet and can be easily broken down by the body to provide quick energy. The smallest carbohydrates are called monosaccharides. *Mono-* is a Greek prefix meaning "one," and saccharide comes from the Greek word *sakcharon*, meaning sugar. A monosaccharide is "one sugar." The single sugars glucose and fructose are monosaccharides.

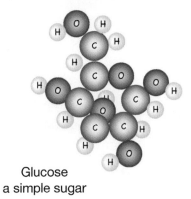

Fructose
a simple sugar

Glucose
a simple sugar

Sucrose is an example of a disaccharide, which is a molecule made of two single sugars. Sucrose contains a molecule of glucose connected by a chemical bond to a molecule of fructose.

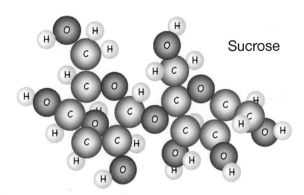

Sucrose

Sucrose is common table sugar and is the same sugar we buy in the store and put on strawberries.

5.5 Starches

When more than a few saccharides, or sugars, are hooked together, the molecule is called a polysaccharide. *Poly* means "many," and a polysaccharide is made of "many sugars." Polysaccharide molecules usually contain ten or more monosaccharides.

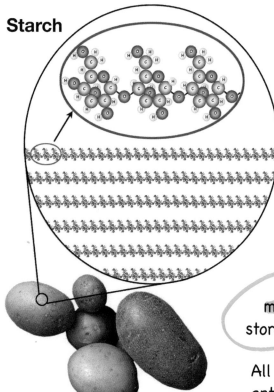

Starch

Potatoes
Courtesy of USDA/
ARS/by Scott Bauer

There are two general types of polysaccharides—starch and cellulose. Starches are the molecules that provide our bodies with most of the energy we need in order to live and work. Potatoes, pasta, and bread are excellent sources of starches.

There are three main kinds of starches. Glycogen is a starch that animals produce in their livers and store in their muscles. Amylose and amylopectin are two starches that are made by plants and are the main energy storage molecules found in rice and potatoes.

All of these polysaccharides are composed entirely of glucose molecules linked together to make long chains. They can have as many as 3,000 glucose units hooked together in a row.

So how do our bodies use these long chains of glucose for energy? Our bodies use special proteins called enzymes to break the long chains of glucose into individual glucose molecules. The single glucose molecules are used directly by the body for energy. But if it is only the glucose our bodies need, why not eat only the simple sugars and have a diet rich in candy-coated sugar bomb cereals?

If we ate only simple sugars, our bodies would use up all of the energy in these molecules too quickly, leaving us feeling tired. The long chains in polysaccharides provide "storage" for the energy molecules so the body can use these over a longer period of time, giving us enough energy to ride bikes, swim, or run.

5.6 Cellulose

Cellulose differs from the starches only in how the glucose molecules are hooked together. The links between the glucose molecules in cellulose are different from the links between those of starches. For the starches, the oxygen atom that connects the two glucose molecules is pointing down. However, the oxygen between the two glucose molecules in cellulose is pointing up. The direction of this bond is the only difference between these two molecules, but it makes a huge difference to us.

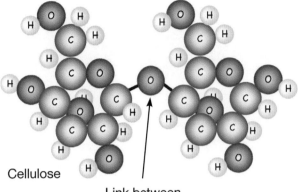
Cellulose

Link between glucose molecules

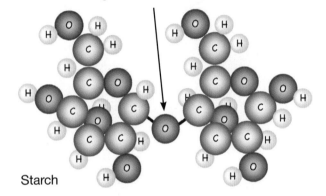
Starch

Cellulose is the main ingredient of wood, cotton, flax, wood pulp, and other plant fibers. It is even in grass. However, wood and grass are not main staples of our diet and are not served for Sunday brunch. In fact, although cellulose has the same glucose molecules that starches have, humans cannot use cellulose for food energy at all.

Many animals, including humans, do not have the enzyme required to break the bonds between the glucose molecules in cellulose. Our enzymes only recognize the bonds in the starches, so we cannot graze the lawn for breakfast. Some animals, like cattle and horses, have bacteria in their digestive system that provide the necessary enzymes to break cellulose links, so these animals can eat grass for food energy, but we cannot.

5.7 Summary

- Some of the nutrients our bodies require to stay healthy are vitamins, proteins, fats, and carbohydrates.

- Carbohydrates are molecules that give our bodies energy.

- Simple carbohydrates are sugars, and larger carbohydrates are the starches and cellulose.

- Long chains of polysaccharides store energy for the body. This energy is released when the body breaks down the chains into individual glucose molecules.

- Our bodies cannot use cellulose for energy because they cannot break the bonds in a cellulose molecule. Animals such as horses and cows can eat grass for food because they have bacteria that provide the enzymes needed to break the bonds in cellulose molecules.

Chapter 6 Technology in Biology

6.1 Introduction	52
6.2 The Botany Laboratory	53
6.3 The Molecular Biology and Genetics Laboratory	**54**
6.4 The Marine Biology Laboratory	56
6.5 Summary	58

A field biologist studying birds
Photo Credit: Travis Booms/USFWS

6.1 Introduction

The study of plants, animals, bacteria and viruses has come a long way from the days of Aristotle. Modern biology is a broad and diverse field of inquiry that includes botany (the study of plants), zoology (the study of animals), cell biology (the study of cells), molecular biology and genetics (the study of biological molecules

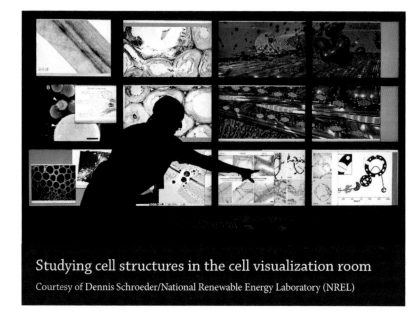

Studying cell structures in the cell visualization room
Courtesy of Dennis Schroeder/National Renewable Energy Laboratory (NREL)

and DNA), anatomy and physiology (the study of animal and plant structure and functions), and many subcategories that include immunology (the study of the immune system), protistology (the study of protists), and marine biology (the study of marine animals and plants), to name a few.

Because biology includes a large collection of specialized categories and subcategories, the technology biologists use can vary greatly from lab to lab. Some biology laboratories may use specialized equipment such as a light microscope or an electron microscope to investigate microscopic cellular structures and molecules. Other laboratories may exist entirely in the field where biologists study large animals, plants, or ecosystems. Some biology labs may look more like chemistry labs, complete with glass beakers, Erlenmeyer flasks, and

Field biologists use maps and electronic devices
Courtesy of Steve Hillebrand/US Fish and Wildlife Service (USFWS)

graduated cylinders. Still other labs may look like military ships with radar equipment, tracking devices, and small submersibles for exploring the deep ocean.

Different types of microscopes are used in many biology labs, and we'll learn more about some of these microscopes in Chapter 7. However, because there is not a typical type of biology lab and because not all biology labs share similar equipment, we'll take a look at just a few different biology labs and the types of instruments they use.

A biologist tracks a black bear
Credit: John & Karen Hollingworth/USFWS

6.2 The Botany Laboratory

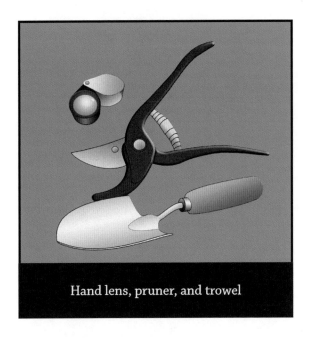

Hand lens, pruner, and trowel

Botany is the study of plants. Biologists who study plants are called botanists. Botanists who do field work may carry a backpack with a variety of different handheld tools such as a trowel for digging into the soil, a pruner for clipping woody samples, a hand lens for examining the small details of plants, a notebook for recording observations, and a compass or GPS for finding the way back to civilization!

Once the botanist has collected samples, the next step might be to take them to a laboratory and use a microscope to examine the fine details of a flower or the cellular structure of the stem or leaf. Dissecting needles may be used to cut leaves, stems, or seeds so their internal structure can be observed. Once the botanist has observed the details of the flower, plant, or leaf, a plant press can be used to preserve the sample for future study.

Floristics is an area of study in botany in which a scientist observes the plants that live in a particular area, noting their number, types, relationships to each other, and how they are distributed throughout the area.

Floristic keys are reference tools used by researchers for identifying plants and plant parts and where the plants live. Floristic keys contain information that may include illustrations of the details of plants and their structure, lists of characteristics used to help identify the plant, seed, or flower being studied, and maps showing where a type of plant grows.

6.3 The Molecular Biology and Genetics Laboratory

Molecular biology and genetics are the study of small molecules and molecular machines inside cells. In addition to the balances, graduated cylinders, and flasks found in many chemistry laboratories, molecular biologists and geneticists often use very specialized equipment. Because many processes inside cells cannot be observed directly with a microscope, molecular biologists and geneticists need to use methods and tools that will allow them to observe what happens inside cells without actually seeing the molecules in action.

Most molecular biology and genetics laboratories use small organisms like bacteria and yeast to study molecules such as proteins, DNA, and RNA, which are the main molecules involved in cell growth, reproduction, and cell death. Many laboratories grow bacterial cultures like *Escherichia coli (E. coli)* to see how they divide and change with the environment. To grow *E. coli*, a laboratory may have an autoclave to sterilize equipment, agar solutions and agar plates for growing and testing *E. coli* cultures, and large warm ovens for maintaining correct growth temperatures.

Proteins, DNA, and RNA can be removed from large volumes of bacterial solutions and analyzed using a method called gel electrophoresis. Gel electrophoresis is a process in which small amounts of proteins, DNA, or RNA are added to a thin gel placed between two glass or plastic plates. Because different sizes of proteins, DNA, and RNA will migrate through the gel at different speeds when an electrical current is passed through the gel, they will separate from one another and can be analyzed.

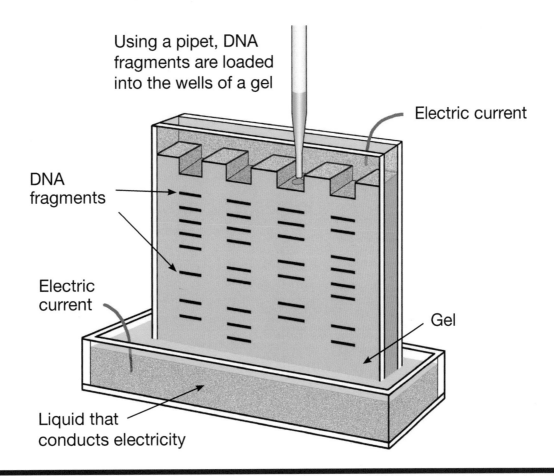

Gel Electrophoresis

1. A sample of DNA is cut into smaller fragments.
2. The DNA fragments are loaded into the wells of a gel that is held between two sheets of glass or plastic.
3. When an electric current is passed through the gel, the DNA fragments move down the gel according to size, with the largest fragments traveling the farthest.

56 Exploring the Building Blocks of Science: Book 6

Filling wells of an electrophoresis gel
Courtesy of Oak Ridge National Laboratory/NREL

Reviewing an electrophoresis gel
Courtesy of Warren Gretz/National Renewable Energy Laboratory (NREL)

6.4 The Marine Biology Laboratory

Marine biology is the study of plants and animals in oceans and other saltwater environments. Marine biologists have floating laboratories on boats or other sea vessels from which they can collect samples and observe sea life. Once back on land, a marine biologist can further analyze the samples in a brick and mortar laboratory.

From a boat, marine biologists may collect samples with the use of a variety of tools and instruments, including water samplers, plankton nets, and sediment corers. With these tools marine biologists can compare water samples, observe the health of small organisms like plankton, test for pollution or other toxins, and take samples of the sediments on the ocean floor along with the organisms that live in the sediments.

Biology—Chapter 6: Technology in Biology   57

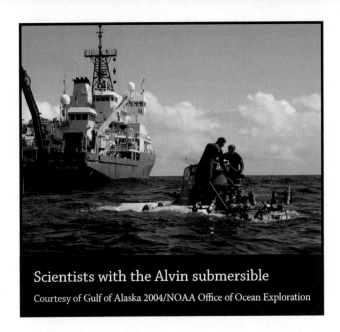

Scientists with the Alvin submersible
Courtesy of Gulf of Alaska 2004/NOAA Office of Ocean Exploration

To observe and collect samples below the surface of the ocean, marine biologists can use scuba equipment, diving bells, submersibles, and remotely operated vehicles (ROV). Scuba equipment allows researchers to swim below the surface for short periods of time. With a diving bell, researchers can travel along the bottom of the ocean for extended periods of time. To go into the deep ocean and travel longer distances, manned submersibles or remotely operated vehicles can be used. By going deeper into the ocean, marine biologists can observe how organisms live in colder and darker water. Scientists believe that the majority of ocean life is still undiscovered.

Marine biologists can also observe the behavior of marine animals by using trackers that send data to satellites. By placing a tracker on a shark, dolphin, or whale, marine biologists can map how the animals move, eat, and reproduce. This is especially helpful for studying sharks to find out when they might be wandering close to a shore full of beachgoers!

Once marine biologists have collected water, plankton, sediment, or other samples from the ocean, they may travel to a land based lab to analyze them. Their lab may have equipment similar to a chemistry or biology lab with balances, graduated cylinders, flasks, and possibly a gel electrophoresis apparatus.

Divers record the diversity of algae
Courtesy of Dr. Jean Kenyon, NOAA/NMFS/PISC/CRED

Collecting botany samples
Courtesy of USFWS

Scientists work with ocean sediment samples
Courtesy of Jeremy Potter/Hidden Ocean 2005 Expedition: NOAA Office of Ocean Exploration

A scientist looks out the window of the Delta submersible
Courtesy of Robert Schwemmer, CINMS, NOAA

6.5 Summary

- Biology is a diverse field of study and includes many different subcategories.

- Because biology includes so many specialized categories and subcategories, the technology biologists use can vary greatly from lab to lab.

- Botanists study plants and may use microscopes, dissecting needles, and plant presses.

- Molecular biologists and geneticists use specialized equipment for processes such as gel electrophoresis to study DNA, RNA, and proteins.

- Marine biologists may have a laboratory on a boat and may use scuba gear, diving bells, submersibles, or remotely operated vehicles to collect samples.

Chapter 7 The Microscope

7.1	Introduction	60
7.2	The Size of Things	61
7.3	The Light Microscope	63
7.4	The Electron Microscope	64
7.5	Scanning Probe Microscopes	67
7.6	Summary	70

Biology

7.1 Introduction

The microscope is an instrument that makes small objects appear bigger. From examining the smallest organisms to mapping the blueprint of the cell, the microscope has dramatically changed the way we understand living things.

The invention of the first microscope is generally credited to Zacharias Janssen, although like many inventions the origin of the microscope is often debated. Zacharias Janssen was born about 1580 CE in Middelburg, the Netherlands. He became a spectacle-maker and developed an expertise in shaping and forming glass lenses. It was widely known at the time that a single, curved glass lens could correct vision and magnify objects. However,

a single lens is limited as to how much it can magnify. In the late 1500s Zacharias and his father Hans Janssen solved this problem by experimenting with combining lenses. They discovered that by using two lenses together the magnification was greatly increased.

The instrument they created resembled a spyglass with two lenses housed in a cylindrical tube. Their instrument became the first primitive compound microscope. It is this advance in technology that eventually led not only to the modern microscope but also the discovery of a fascinating new world of microscopic biology.

About a century later, a Dutch lens maker named Anton van Leeuwenhoek (1632-1723 CE) perfected the polishing and grinding of lenses. Although he probably did not combine lenses to make a compound microscope, one of his lenses could

---

Previous page—Scanning electron microscope image credits: 1. Bug eye, CDC/Janice Carr; 2. Exoskeleton surface of a mite, CDC/William L. Nicholson, PhD, Cal Welbourn, PhD, Gary R. Mullen; 3. Mosquito Head, CDC/Paul Howell

magnify samples up to 300 times. Leeuwenhoek was the first person to observe tiny creatures in pond water, bacteria, and red blood cells. Around the same time, Robert Hooke (1635-1703 CE), an English scientist, improved upon the Janssen microscope and was able to see the outline of cells in thinly sliced pieces of cork. Hooke became well known for his book *Micrographia* in which he wrote about and illustrated his observations.

Today, there are several different kinds of microscopes. The kind of microscope the Janssens invented and Hooke improved on is called a light microscope. A light microscope uses light and the interaction of light with glass lenses to focus and magnify an object. In addition to the light microscope, scientists can now use electrons and probes to image small objects. The electron microscope uses a beam of electrons to magnify objects, and a class of microscopes called probe microscopes use a stylus, or small probe, to "feel" the features on the surface of a sample. We will learn more about light, electron, and probe microscopes later in this chapter.

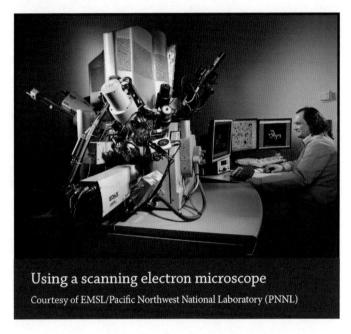

Using a scanning electron microscope
Courtesy of EMSL/Pacific Northwest National Laboratory (PNNL)

7.2 The Size of Things

How big is an insect eye, a piece of hair, a protist, or a red blood cell? We can see a fly sitting on a wall, but with our naked eyes can we see mitochondria or a cell from the inside of our cheek? What size are the smallest things the human eye can see, and what size are the smallest things microscopes can help us see?

Before we take a look at different microscopes, it's important to explore the sizes of objects and the microscopic scale. The smallest object the unaided human eye can resolve is about the size of a strand of hair. Anything much smaller than a strand of hair needs to be magnified to be seen.

A Note About Units

The metric scale is the preferred unit for measuring the size of objects with a microscope. Although metric units can be converted to British units, the British scale is never used as a measurement for microscopic objects. In the metric scale the meter is the base unit used to measure length. Units larger or smaller than the meter are multiples of ten and are given a prefix attached to the word *meter* to identify them. For example, a length 100 times smaller than the meter is called the centimeter. A length 1000 times smaller than the meter is called a millimeter. The following table lists the metric units most commonly used in microscopy.

Metric Measurement	Number of Times Smaller than a Meter	Measurement in Meters
1 centimeter (cm)	10^{-2} = 1 hundred times	0.01 meter
1 millimeter (mm)	10^{-3} = 1 thousand times	0.001 meter
1 micrometer (um)	10^{-6} = 1 million times	0.000001 meter
1 nanometer (nm)	10^{-9} = 1 billion times	0.000000001 meter

Insect Wing	Strand of Hair	Protist	Plant Cell	Animal Cell	Red Blood Cell	Bacterium	Mitochondrion	Lysosome
6 mm	400 um	200 um	100 um	20 um	8 um	2 um	500 nm	200 nm

Note: These sizes are approximate. Sizes of actual objects will vary.

Red blood cell
Courtesy of CDC/
Janice Haney Carr

A strand of hair is around 0.4 mm (400 um) and is near the limit of what can be seen by the unaided human eye. Protists are around 200 um, plant cells are around 100 um, animals cells are around 20 um, and red blood cells are around 8 um. An *E. coli* bacterium is around 2 um (2000 nm), and most structures inside a cell are 200 nm and below.

7.3 The Light Microscope

The basic compound light microscope consists of an eyepiece or ocular lens, the magnifying (objective) lens, a stage for holding the sample, a light source, and various areas for adjusting focus, brightness, and magnification.

A light microscope works when light is passed through a sample and the lenses collect the light, bend the light to provide magnification, and separate the bent light to make details in the sample visible. In other words, a microscope magnifies a sample, resolves details in a sample, and creates contrast so a sample can be seen.

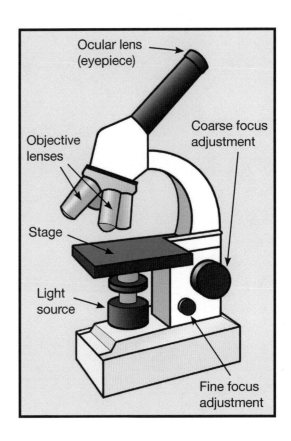

Objective lenses on a light microscope
Courtesy of Morgan Brown/CSIRO

The magnification of a lens (how much a microscope can enlarge a sample) is expressed as numerical multipliers. A 2X magnification means that the lens doubles the size of the sample. In a 10X magnification the lens makes the sample 10 times larger, and a 100X magnification makes the sample 100 times larger. A modern compound microscope can magnify an object up to 1000 times!

Both the eyepiece and the objective lens magnify a sample. The eyepiece is the lens you look through, and the objective lens is the lens closest to the sample. The total magnification is the magnification of the eyepiece times the magnification of the objective lens. For example, if the eyepiece is 10X and the objective lens is 40X, the total magnification is 400X. The very smallest objects the unaided human eye can see are around 0.1 millimeter (100 micrometers), so objects smaller than 100 micrometers need to be enlarged to be seen. This is what magnification does—it enlarges small objects so we can both see them and observe details.

However, magnification of a sample is not the only important feature of a microscope. More important than magnification is the resolution of the sample by the lens. Resolution is the smallest distance between two points on a sample that a lens can clearly define as being separate. Resolution is essentially the ability of a microscope to separate fine details on a sample.

For example, imagine you have two dots on a page. If we magnify the dots, we can make them appear bigger, but if we cannot separate them, they will appear to be a single, merged dot.

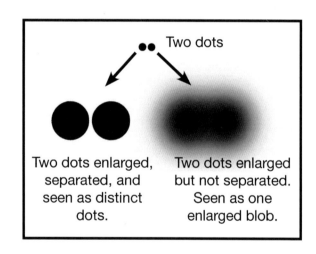

Two dots enlarged, separated, and seen as distinct dots.

Two dots enlarged but not separated. Seen as one enlarged blob.

The resolution depends on the wavelength of light and the quality of the objective lens. The objective lens is the lens closest to the sample, and it is the only lens that is involved in providing resolution. The ocular lens (eyepiece) only magnifies what the objective lens resolves. In other words, if you use a higher magnification on the eyepiece but the objective lens cannot separate two points, the image will look big but will be blurry.

7.4 The Electron Microscope

Light microscopy is limited by the wavelength of visible light. Objects below about 300 nm cannot be resolved with a light microscope. This means that many small structures inside and on the surface of cells cannot be observed. Before the 1930s many scientists believed that it would never be possible to see things that are smaller than 300 nm. However, during the 1930s scientists began experimenting with electrons, and by the late 1950s the electron microscope had broken through the resolution limit of the light microscope.

The electron microscope uses a beam of electrons focused by a magnetic field much in the way a glass lens focuses a light beam. The "lens" for an electron microscope is called a solenoid which is a coil of wire wrapped around the outside of a tube. When an electric current is passed through the wire, an electromagnetic field is created that can be used to control the electron beam.

The resolution of the electron microscope depends on how fast the electrons are traveling. The faster the electrons travel the shorter their wavelength and the higher the resolution. With an electron microscope, samples can be magnified almost 2 million times!

Electron microscopes are very expensive pieces of equipment and are generally found only in specialized research labs. The electron microscope has a large vacuum chamber that houses an electron gun and magnetic lenses. The sample is inserted in the viewing chamber which is below the vacuum chamber, and a beam of electrons is directed from the electron gun toward the sample below. In a scanning electron microscope (SEM), an electron beam is scanned across the sample surface, and a detector picks up the electrons that have been scattered during the scan. A computer then uses this information to create an image.

Diagram of a scanning electron microscope

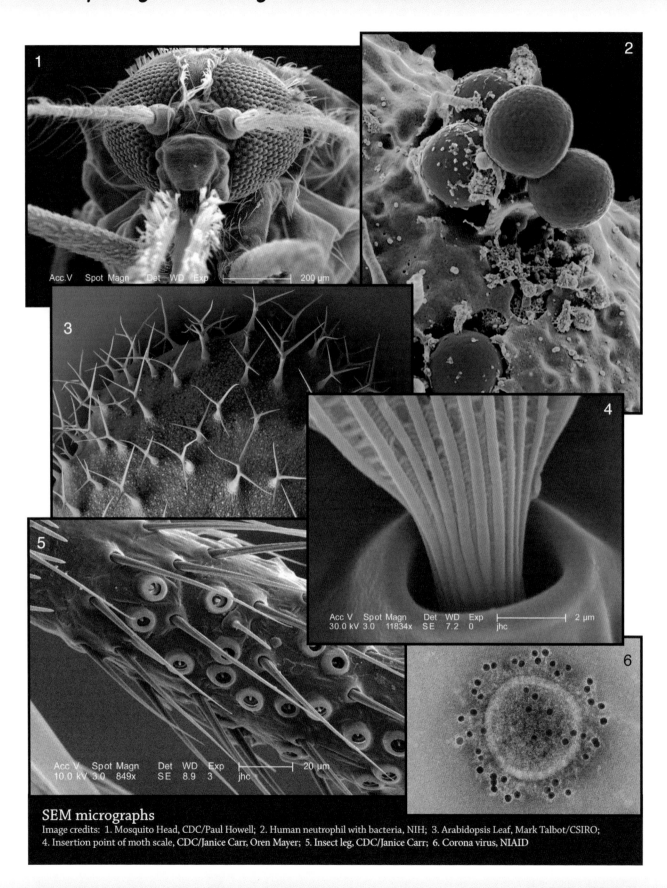

SEM micrographs
Image credits: 1. Mosquito Head, CDC/Paul Howell; 2. Human neutrophil with bacteria, NIH; 3. Arabidopsis Leaf, Mark Talbot/CSIRO; 4. Insertion point of moth scale, CDC/Janice Carr, Oren Mayer; 5. Insect leg, CDC/Janice Carr; 6. Corona virus, NIAID

7.5 Scanning Probe Microscopes

How small are molecules? How small are atoms? How small is DNA, RNA, or a protein? The cells that make up living things are small, but even the smallest living cell is made up of billions of proteins, molecules, and atoms!

For a long time scientists could not observe small molecules found in cells or the atoms that make them. However, in the 1980s a new microscope technology was invented. This new device is called a **scanning tunneling microscope**, or STM. An STM is part of a family of microscopes called **probe microscopes** that make it possible to "see" small molecules and even atoms.

A scanning tunneling microscope is not a typical microscope. It does not work with light or lenses, and you don't look through it. In fact, when using an STM, you do not actually "see" the atoms, at least not in the way that you are looking at this page in front of you.

An STM works by "scanning" the surface of an object and then projecting an image of this surface onto a computer monitor or other screen. The STM has a metal **probe** called a stylus that actually does the scanning. The stylus is extremely sharp—it comes to a point that is only one atom wide!

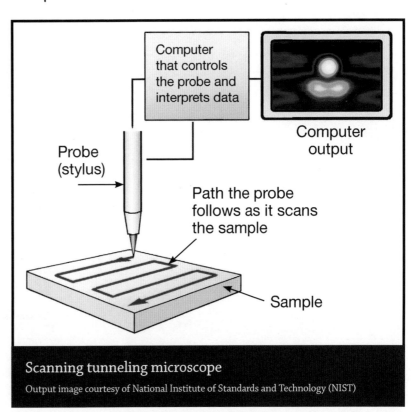

Scanning tunneling microscope
Output image courtesy of National Institute of Standards and Technology (NIST)

The stylus is controlled by a computer and moves back and forth over the surface of the object that is being scanned. The stylus stays very close to the surface of the object with the gap between the tip of the stylus and the object being about as wide as one atom, or even smaller. The precision required to keep the stylus moving at the right distance from the scanning surface would not be possible without computers. As the stylus moves, it "picks up" electrons from the surface of the object. The electrons show where the atoms in the object are located. The STM electronically amplifies the signals created by these electrons and a computer then interprets the signals, creating an image on a monitor.

Xenon atoms written on a surface to spell IBM
Reprinted with permission from IBM Corporate Archives

An STM can produce phenomenal images of a surface, but it has another amazing function. An STM can be used to "grab" individual atoms! The computer controlling the STM can then move the atoms to specific locations.

In 1990, researchers at IBM used an STM to grab individual xenon atoms. It took over 20 hours, but they were able to arrange 35 atoms into the letters I, B, and M to make the smallest company logo ever.

Since then, researchers have been discovering ways to move atoms around more quickly and how to make incredibly tiny structures, one atom at a time.

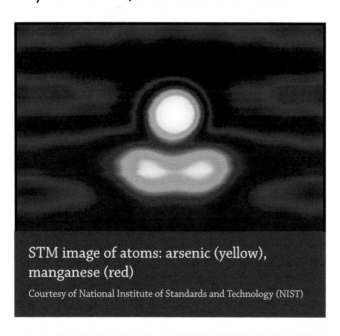

STM image of atoms: arsenic (yellow), manganese (red)
Courtesy of National Institute of Standards and Technology (NIST)

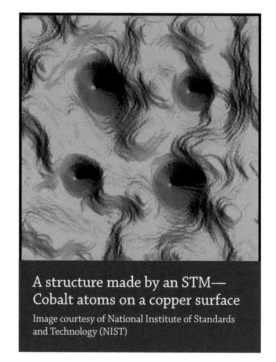

A structure made by an STM—Cobalt atoms on a copper surface
Image courtesy of National Institute of Standards and Technology (NIST)

One of the drawbacks of the early scanning tunneling microscopes was that they could only be used to scan objects such as metals that conduct electricity easily. Therefore, they could not be used to create images of substances that were not conductors of electricity, such as plastics or living tissues. In the years since STMs were invented, several other types of probe microscopes have been developed. They work in slightly different ways, but all of these microscopes allow scientists to get an extremely close-up image of very small objects.

One of these probe microscopes is the atomic force microscope (AFM) which can scan many different types of surfaces, including metals and nonmetals. Like an STM, an AFM stylus has a very sharp tip. But instead of picking up electrons to create an image like an STM does, an AFM can "see" atoms by just bumping into them (that is, by measuring the *force* between an atom and the tip of the probe).

Because everything is made of atoms, an AFM can "see" all kinds of materials, not just those that conduct electricity. The AFM has been used to image the surfaces of cells and observe small proteins in action.

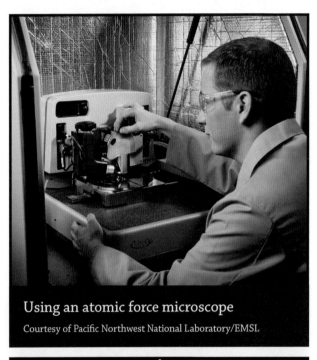

Using an atomic force microscope
Courtesy of Pacific Northwest National Laboratory/EMSL

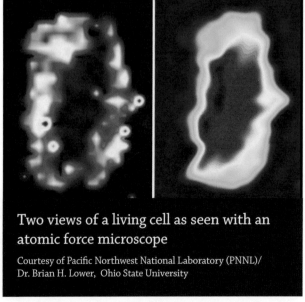

Two views of a living cell as seen with an atomic force microscope
Courtesy of Pacific Northwest National Laboratory (PNNL)/
Dr. Brian H. Lower, Ohio State University

7.6 Summary

- Different types of microscopes include the light microscope, the electron microscope, and a family of probe microscopes.

- A light microscope works when light passes through a sample and lenses collect the light, provide magnification of the sample by bending the light, and separate the bent light so that details in the sample are visible.

- The amount of magnification of a light microscope is the magnification of the ocular lens (eyepiece) times the magnification of the objective lens (the lens closest to the sample).

- The objective lens determines the resolution of a light microscope.

- Electron microscopes use a beam of electrons and a magnetic "lens" to image a sample.

- Probe microscopes, including the scanning tunneling microscope (STM) and atomic force microscope (AFM), scan the surface of a sample and can produce images of small molecules and atoms.

Chapter 8 Protists

8.1 Introduction   72

8.2 Classification   73

8.3 Photosynthetic Protists   76

8.4 Heterotrophic Protists   78

8.5 Summary   82

Biology

Photo credits: See page 82

8.1 Introduction

Protists, sometimes called protozoa, are organisms that are like both plants and animals. Protists are in the domain Eukarya and have their own kingdom called Protista. The word Protista comes from the Greek *protos* which means "first." Although it is not likely that protists were among the very first life forms to appear on the planet, they are some of the oldest organisms that have been found in the fossil records.

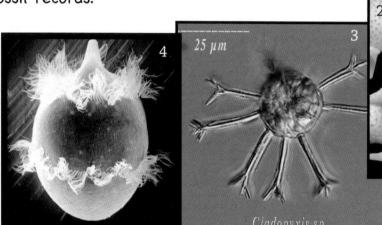

Protists come in a variety of shapes and sizes. Most protists are too small to see with the naked eye. Many protists are made of only one cell, but some protists, like kelp and seaweed, group together into large colonies. For most of human history nobody knew protists existed. However, when the first microscopes were invented in the middle of the 17th century, an entirely new world of microscopic organisms, including protists, was found. Protists live almost everywhere, including soil, freshwater ponds, and saltwater oceans.

1. Dinoflagellate, Courtesy of CSIRO; 2. Formanifora, Courtesy of Psammophile (CC BY SA 3.0); 3. Dinoflagellate, Courtesy of Dr. John R. Dolan, Laboratoire d'Oceanographique de Villefranche; Observatoire Oceanologique de Villefrance-sur-Mer; 4. *Didinium nasutum*, Courtesy of Gregory Anitpa, San Francisco State University; 5. Giant kelp, Courtesy of Claire Fackler, CINMS/ NOAA

Biology—Chapter 8: Protists

8.2 Classification

It is unknown how many species of protists exist, but estimates range from 36,000 to 200,000, many of which have not yet been discovered. Although protists are classified in the single kingdom, Protista, they vary in structure and function more than any other group of organisms. Because this group is so diverse, there are several different classification systems for protists. In this text we will focus on four main groups depending mostly on how they move. These group are the ciliates, the flagellates, the amoebas, and the spore-forming protists.

The ciliates are in the phylum Ciliophora. Ciliophora swim by using cilia, which are very small hair-like projections on their bodies. The cilia beat very rapidly to propel the organism through the water like a little submarine. The ciliates include *Paramecium*, *Didinium*, *Hypocoma*, and *Stentor*.

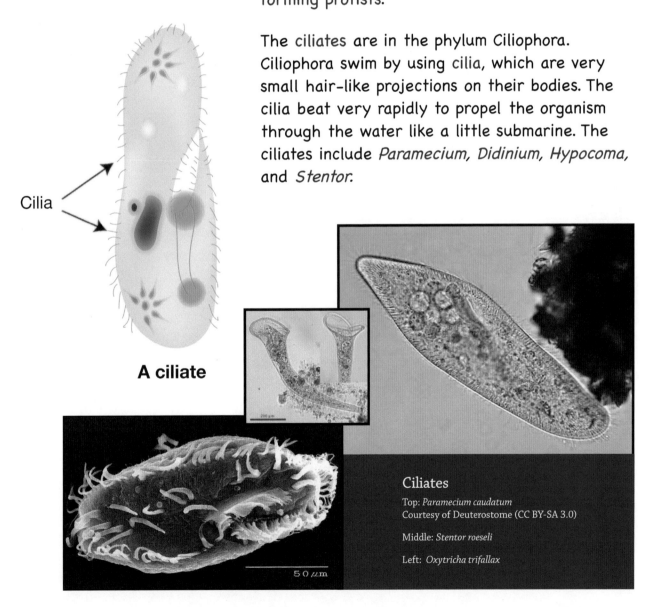

Cilia

A ciliate

Ciliates
Top: *Paramecium caudatum*
Courtesy of Deuterostome (CC BY-SA 3.0)

Middle: *Stentor roeseli*

Left: *Oxytricha trifallax*

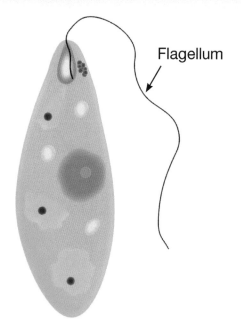

A Flagellate

There are several different phyla for the flagellates including Trichozoa, Euglenozoa, Dinozoa, Choanozoa, and Metamonada. Flagellates also swim, but instead of many short, hair-like projections, flagellates have one, or more long, whip-like flagella that extend from one end of their body. These whips propel the flagellates through the water much like the tail of a fish. Many flagellates have a thin outer covering called a pellicle. Flagellates can exist as single organisms or in colonies. Many flagellates are parasitic, living inside other organisms.

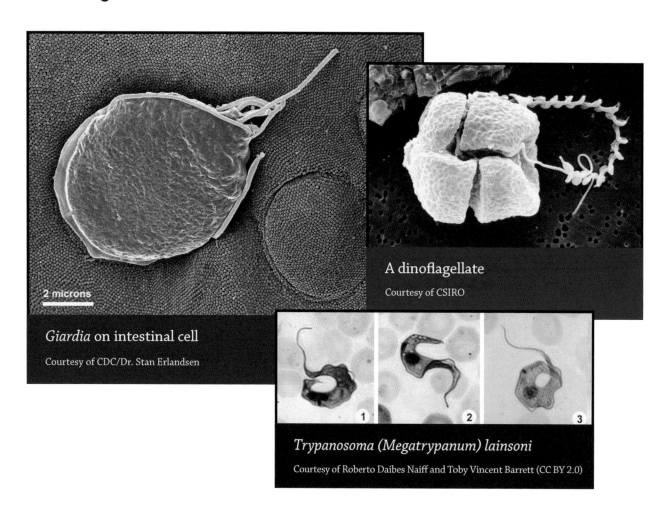

Giardia on intestinal cell
Courtesy of CDC/Dr. Stan Erlandsen

A dinoflagellate
Courtesy of CSIRO

Trypanosoma (Megatrypanum) lainsoni
Courtesy of Roberto Daibes Naiff and Toby Vincent Barrett (CC BY 2.0)

Biology—Chapter 8: Protists

Amoeba

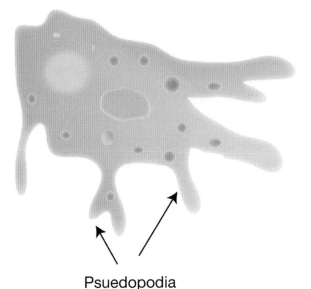

Psuedopodia

Amoebas move very differently than the ciliates and flagellates. Amoebas do not swim or use flagella or cilia but instead crawl along surfaces by extending and bulging the edges of their membranes. The portions of their membranes that stick out are called pseudopodia. *Pseudo* is Greek and means "false" and *podia* means "feet," so pseudopodia are "false feet." Once the pseudopodia are extended, the rest of the amoeba flows into them, pulling the amoeba forward. The process then begins again.

In a microscope, the movement of an amoeba along the surface of a glass slide looks something like the following illustration.

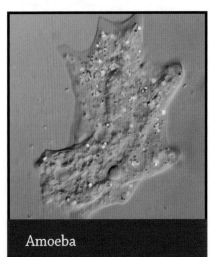

Amoeba

Courtesy of Picturepest (original picture), Nina-marta (modification) - CC BY 2.0

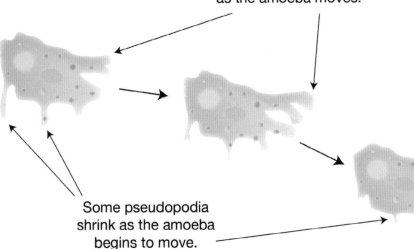

Some pseudopodia extend as the amoeba moves.

Some pseudopodia shrink as the amoeba begins to move.

Sporozoa, or spore-forming protists, include three major phyla—Apicomplexa, Microspora, and Myxosporidia (Myxospora). Sporozoans live as parasites within cells or organs of almost every kind of animal. Sporozoans do not have flagella, cilia, pseudopodia, or any other locomotion process. A Sporozoan spends much of its life cycle unable to move by itself and passes from host to host in a protective capsule called a spore.

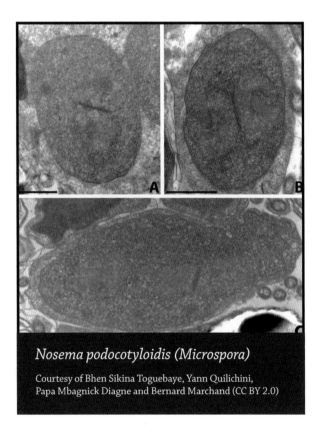

Nosema podocotyloidis (Microspora)

Courtesy of Bhen Sikina Toguebaye, Yann Quilichini, Papa Mbagnick Diagne and Bernard Marchand (CC BY 2.0)

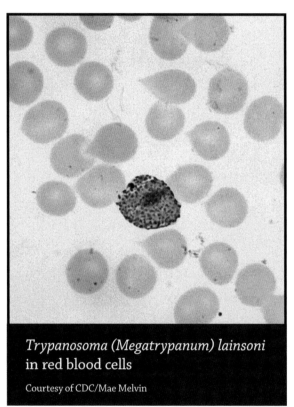

Trypanosoma (Megatrypanum) lainsoni in red blood cells

Courtesy of CDC/Mae Melvin

8.3 Photosynthetic Protists

Because most protists are single-celled organisms, they do not have the advantage of using tissues and organs to process food. Instead, they must gather food, digest nutrients, and eliminate wastes, all within a single cell. As a result, protists are much more complicated than cells of other eukaryotic organisms.

Some protists contain chloroplasts and use carbon dioxide, water, and the Sun's energy to make food by photosynthesis, just like plants do. Organisms that make

Biology—Chapter 8: Protists

their own food are called autotrophs. Autotroph comes from the Greek words *auto* which means "self" and *trophe* which means "food or nourishment," so an autotroph is an organism that is fed by itself.

Euglena viridis is an example of a photosynthetic protist. *Euglena* are found in freshwater streams and ponds, sometimes being so numerous that the water turns green. Because *Euglena* depend on photosynthesis for food, it is important for them to be able to detect the sunny areas in a pond or stream. To detect light, *Euglena* have a small red spot toward the end of their body near the flagellum. This spot is called the eyespot or stigma. The stigma is a light sensitive area shaped like a shallow cup. This shape allows the *Euglena* to detect sunlight only from a particular direction. When the *Euglena* is traveling toward the light, a small part in the base of the stigma is illuminated. When the *Euglena* swims away from the light, the spot is no longer illuminated, and the *Euglena* knows that it is no longer in the path of the sunlight. Using the stigma as a detector, the *Euglena* can find the sunlight needed for photosynthesis.

Euglena sp. (species)
Courtesy of Deuterostome (CC by SA 3.0)

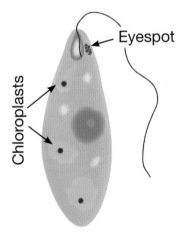

Euglena viridis

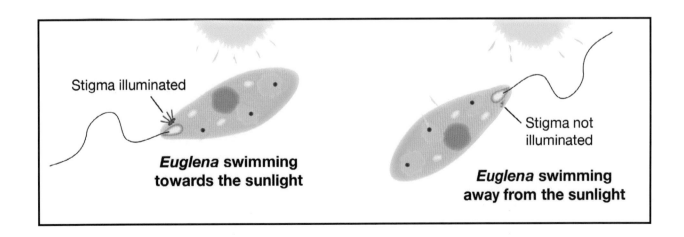

Euglena swimming towards the sunlight

Euglena swimming away from the sunlight

8.4 Heterotrophic Protists

Many protists do not have the ability to make their own food through photosynthesis. They need to eat, just like we do. Organisms that cannot make their own food are called heterotrophs. Heterotroph comes from the Greek words *hetero* which means "other or different" and *trophe* which means "food or nourishment," so heterotrophs need to find food from sources other than themselves.

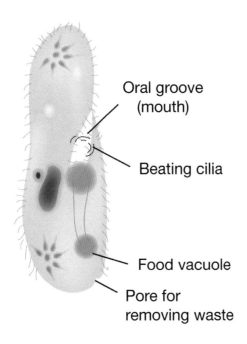

Paramecia, for example, live on bacteria, algae, and other small organisms. They have an oral groove that acts just like a big mouth. They gather their food by rapidly beating the cilia near the oral groove and creating water currents that sweep the food into the opening. The food travels into a food vacuole, which is like a tiny stomach for the *Paramecium*. Once food is inside, the vacuole circulates around the cell as the food is being digested. Any undigested food left in the food vacuole is ejected through a small pore.

Paramecium

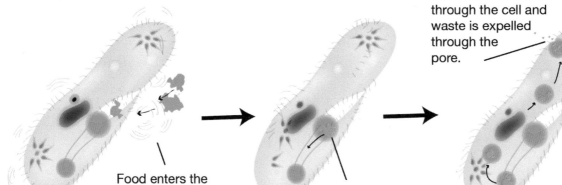

Food enters the oral groove as the cilia swirl the surrounding water.

Food in the oral groove enters and travels to the vacuole.

The food vacuole travels through the cell and waste is expelled through the pore.

The food enters the food vacuole and gets moved around the cell.

Amoebas are another type of protist that cannot make their own food. Amoebas are hunters; they feed on algae, other protozoa, and even other amoebas. However, an amoeba won't eat everything that comes its way. It is a picky eater.

An amoeba does not have a cellular mouth like a paramecium. It can swallow food anywhere on its body. When an amoeba encounters something tasty, it thrusts its pseudopodia outward to surround the prey.

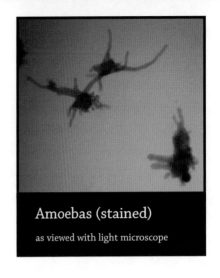

Amoebas (stained) as viewed with light microscope

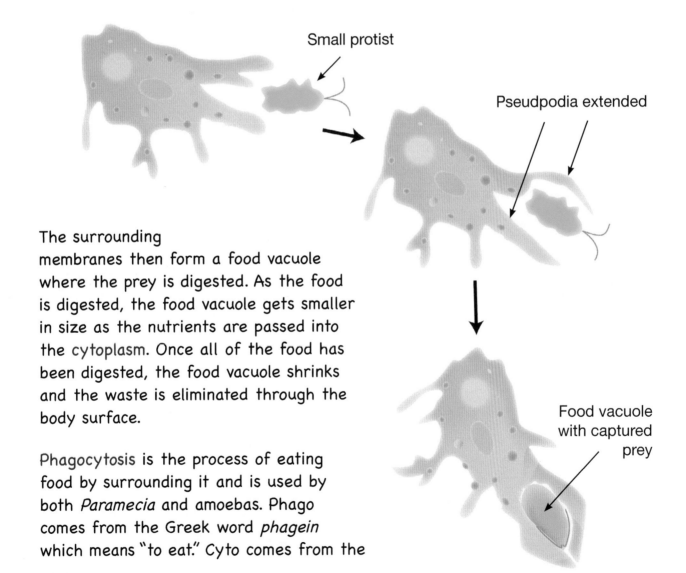

The surrounding membranes then form a food vacuole where the prey is digested. As the food is digested, the food vacuole gets smaller in size as the nutrients are passed into the cytoplasm. Once all of the food has been digested, the food vacuole shrinks and the waste is eliminated through the body surface.

Phagocytosis is the process of eating food by surrounding it and is used by both *Paramecia* and amoebas. Phago comes from the Greek word *phagein* which means "to eat." Cyto comes from the

Greek *kytos* which means a receptacle or container. The word-forming element cyt- is used by biologists to refer to a cell, so phagocyte is "a cell that eats."

There are still other protists that use entirely different methods for capturing and consuming food. *Didinium*, for example, has a single small tentacle called a toxicyst which contains a substance that is poisonous to *Paramecia*. A *Didinium* pierces a *Paramecium* with the toxicyst to paralyze it and then swallows the *Paramecium* whole. *Didinia* are barrel shaped and have bands of cilia around their body, allowing them to swim fast and move in different directions.

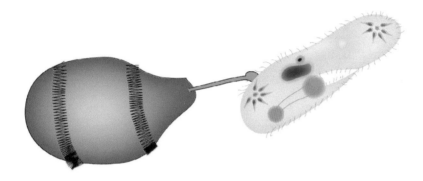

Didinium

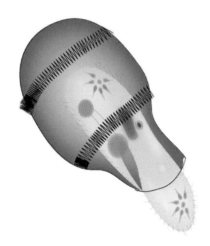

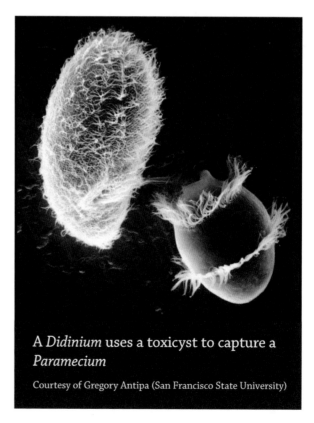

A *Didinium* uses a toxicyst to capture a *Paramecium*

Courtesy of Gregory Antipa (San Francisco State University)

Podophrya, on the other hand, have many tentacles with knobbed ends. A *Podophrya* begins its life as a free-swimming ciliate. When it matures, it loses its cilia, grows tentacles, and uses a stalk to attach itself to a surface. When a protist is swimming past, the *Podophrya* bends and moves to try to capture the prey. If the passing protist touches a tentacle, it sticks to the tentacle and becomes paralyzed. The *Podophrya* then uses its tentacle to put enzymes into the captured protozoa to break it down into molecules that can be absorbed by the *Podophrya* for food.

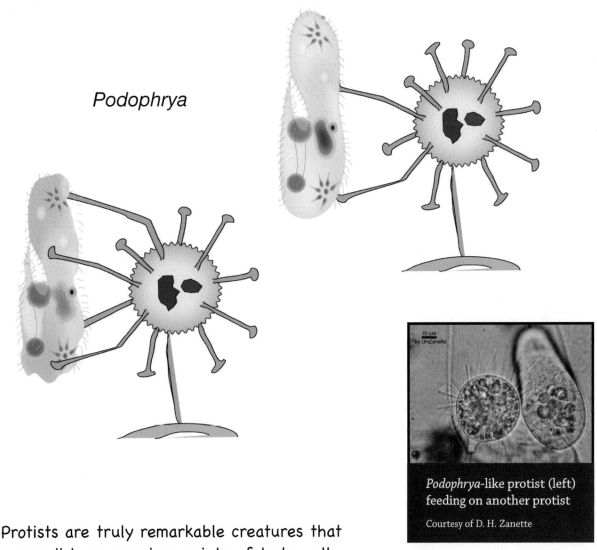

Podophrya

Podophrya-like protist (left) feeding on another protist

Courtesy of D. H. Zanette

Protists are truly remarkable creatures that accomplish an amazing variety of tasks—all within a single cell!

8.5 Summary

- Protists are microscopic, one-celled organisms that have both plant-like and animal-like qualities.

- There are four main types of protists that are classified primarily on how they move. These are ciliates, flagellates, amoebas, and Sporozoans.

- Ciliates move with tiny hair-like projections called cilia.

- Flagellates move with one or more long whip-like structures called flagella.

- Amoebas move by crawling with pseudopodia, or "false feet."

- Photosynthetic protists (autotrophs), such as *Euglena*, use the Sun's energy to make food.

- Heterotrophic protists, including *Paramecia*, amoebas, *Didinium*, and *Podophrya* capture other organisms for food by using cilia, pseudopods, or tentacles.

Image credits for page 71 (chapter title page):

1. Giant kelp, Courtesy of Claire Fackler, CINMS/ NOAA
2. Formanifora, Courtesy of Psammophile (CC BY SA 3.0)
3. Star Radiolarian, Courtesy of Dr. John R. Dolan, Laboratoire d'Oceanographique de Villefranche; Observatoire Oceanologique de Villefrance-sur-Mer
4. *Giardia* on intestinal cell, Courtesy of CDC/Dr. Stan Erlandsen
5. *Giardia*, Courtesy of CDC/Dr. Stan Erlandsen
6. *Didinium nasutum* eating a *Paramecium*, Courtesy of Gregory Antipa (San Francisco State University) and H. S. Wessenberg (San Francisco State University)

Chapter 9 Fungi: Molds, Yeasts, Mushrooms

9.1	Introduction	84
9.2	Classification of Fungi	85
9.3	Structure of Fungi	86
9.4	Reproduction of Fungi	87
9.5	Phylum Zygomycota-Molds	87
9.6	Phylum Ascomycota-Yeasts, Truffles	89
9.7	Phylum Basidiomycota-Mushrooms	92
9.8	Summary	94

9.1 Introduction

What grows in the dirt like a plant but is not green and isn't a plant? Mushrooms! Although mushrooms grow in the dirt like plants and at one time scientists grouped mushrooms with plants, we now know that mushrooms are not plants. Mushrooms are one type of organism in a group of living things called fungi.

Fungi is the plural form of the word fungus which is the Latin word for mushroom. Fungi are in the domain Eukarya and have their own kingdom called Fungi that is separate from plants, animals, and protists. The study of fungi is called mycology which comes from the Greek word *mykes* which means "slippery." If you have ever touched the top of a mushroom or felt the mold on bread, you might remember that it felt slippery.

There are around 100,000 known species of fungi that include yeasts, mildews, molds, and mushrooms, and it is thought that there are many more species that have not yet been discovered. Fungi are found in large numbers in many places on Earth including lakes, soils, the air, rivers, and oceans. We use some fungi for food. For example, many different

Photo credits: 1. and 2., USFWS; 3. Mold in a petri dish, Scott Bauer, USDA/ARS
4. Jennifer Winston, USFWS; 5. Chelsi Hornbaker, USFWS; 6. M Murrill Fisher, USDA Forest Service; 7. Steve Hillebrand, USFWS

types of non-toxic mushrooms and truffles are eaten in soups, stews, and on pizza! Other fungi, such as yeast and molds, are used in bread, cheese, and beer making.

Fungi come in a variety of different shapes, sizes, and colors. Many of the brilliantly colored mushrooms are beautiful to see but are poisonous if eaten. The smallest fungi, such as yeasts, are microscopic and have only one cell, while other fungi are multicellular. Some fungi are very large. The Phellinus ellipsoideus found in China can form a massive fruiting body (the part of a fungus that produces spores for reproduction). One Phellinus ellipsoideus found on the Hainan Island in southern China, measured 10.8 meters (35.4 ft.) long and almost a meter (3 ft.) wide.

Meripilus giganteus (phylum Basidiomycota)
Courtesy of Axel Kuhlmann

Fungi are not mobile, like animals are, and unlike plants, fungi do not make their own food. Fungi get food from other organisms, some dead and some alive. Many fungi are saprophytes, which means that they use dead or decaying matter for food. Other fungi are parasites and feed on living things. Soils that contain an abundance of organic matter are an excellent environment for fungi, and many species can be found on the forest floor and growing on decaying trees.

9.2 Classification of Fungi

Fungi were originally classified in the plant kingdom, but we now know that fungi don't have chlorophyll and don't produce their own food; therefore, they are very different from plants. The kingdom Fungi is divided into three, four, six, or more phyla depending on the way different characteristics are classified. Some scientists divide the kingdom into three phyla based on how fungi reproduce. Other scientists classify fungi based on their molecular biology.

In this text we will explore the three phyla Zygomycota, Ascomycota, and Basidiomycota which are defined by how the fungi reproduce.

1, 2. Grapefruit mold & mold cultures—Courtesy of Scott Bauer, USDA/ARS; 3. Morel; 4. Aleuria Aurantia; 5. Candida yeast—Courtesy of CDC/Melissa Brower (artist's rendition); 6. Chicken of the Woods—Courtesy of National Park Service; 7. Bracket fungi—Courtesy of USFWS/Chelsi Hornbaker; 8 & 9. Mushrooms with caps; 10. Mushroom showing gills—Courtesy of National Park Service

9.3 Structure of Fungi

With the exception of yeasts, the main part of a fungus is called the mycelium. The mycelium is a mass of tiny, thread-like structures called hyphae (singular, hypha) that resemble the roots of plants. Because the hyphae are so numerous, they provide the fungus with a large amount of surface area for absorbing nutrients. Since hyphae can absorb only small molecules, they first release digestive enzymes to break down the organic matter that surrounds them. The hyphae then absorb the smaller molecules for the fungus to use for food. Each hypha is surrounded by a tough outer cell wall made of chitin, which is the same carbohydrate that insects use to build their exoskeletons.

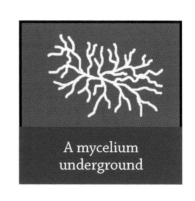

A mycelium underground

9.4 Reproduction of Fungi

Micrograph of yeast budding
Courtesy of CDC/Dr. Libero Ajello

Fungi can reproduce both asexually and sexually. In asexual reproduction an organism reproduces itself without combining its genetic material with the genetic material of another organism. Asexual reproduction occurs in yeast and some other fungi in the form of budding or fission when one cell divides into two or more cells. All the resulting cells will have the same genetic makeup as the original cell. Another way many fungi can reproduce asexually occurs when a small piece of a hypha breaks off and starts growing into a new organism, or mycelium.

Each of the different fungi phyla presented in this chapter has slightly different ways of reproducing sexually. Sexual reproduction involves the sharing of genetic material between two organisms and is more complicated than asexual reproduction. Fungi do not have male and female organisms. Instead they have different mating types with each mating type containing different genetic material. Some species of fungi have two mating types, while other species have more than two. In general, sexual reproduction in fungi occurs when hyphae of two different mating types touch and exchange genetic material. The hyphal cells then merge the genetic material and make more cells through a series of specific steps in which the merged genetic material is copied. When a new mycelium is formed, it will have the merged genetic material which is different from the genetic material of either of the parent mycelia. This produces a new strain of the fungus. A strain is a group of organisms that have the same genetic makeup as each other.

9.5 Phylum Zygomycota—Molds

Zygomycota is the phylum that contains molds. Since moldy food is not safe to eat and tastes bad, we often think of molds as our enemy. But molds, along with other fungi and bacteria, are necessary for breaking down dead plant material and turning it into soil. If this plant material were not broken down, or decomposed, by fungi and bacteria, living plants would not be able to get the nutrients they need to grow.

Some of the organisms in the phylum Zygomycota are among the fungi that form mycorrhizae, which are symbiotic relationships between fungi in the soil and plant roots. A symbiotic relationship is one that is beneficial to both organisms. In a mycorrhiza, the fungus growing on a plant's roots sends out hyphae that spread farther out into the soil. Some of he nutrients the fungus absorbs become available to the plant, giving the plant access to nutrients that are beyond the reach of the plant's root

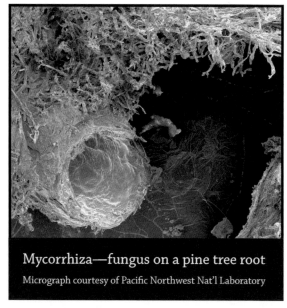

Mycorrhiza—fungus on a pine tree root
Micrograph courtesy of Pacific Northwest Nat'l Laboratory

system. The amount of nutrients available to the plant increases, and the plant in turn provides the fungus with nutrients that the fungus needs to grow and function, such as sugars that the plant makes through photosynthesis.

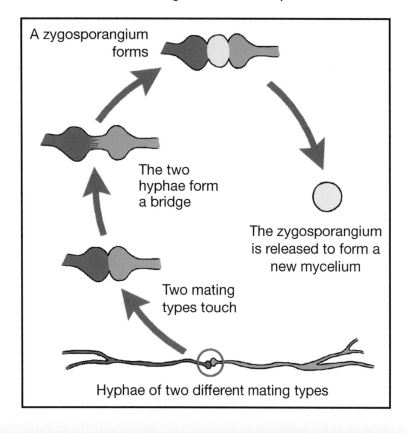

The zygomycetes (members of the phylum Zygomycota) have different mating types for sexual reproduction. After the hyphae of two zygomycetes of different mating types touch, they form a bridge between them. Genetic material from each of the hyphae goes into the bridge and is merged. The bridge with the merged genetic material grows into a new cell called a zygosporangium. The zygosporangium is a spore

Biology—Chapter 9: Fungi: Molds, Yeasts, and Mushrooms   89

from which a new mycelium can grow. Since the zygosporagium contains merged genetic material from two different mating types, the new organism will have different genetic material from either of the original mating types.

When conditions are unfavorable for growth, the zygosporangium rests, or doesn't grow. The zygosporangium has a tough outer coat to protect it and is able to survive harsh conditions such as freezing or drying. When conditions improve, the zygosporangium can germinate and become a new organism. The phylum Zygomycota is named for the tough zygosporangium.

Zygomycetes can make spores asexually (without having the two mating types come together to share genetic material) by forming a structure called a sporangium where spores are produced. A sporangium forms at the tip of a hypha and looks like a ball mounted on a stem—something like a tree from a Dr. Seuss book! When the spores are mature, they are released from the sporangium. All of the spores will have the same genetic makeup.

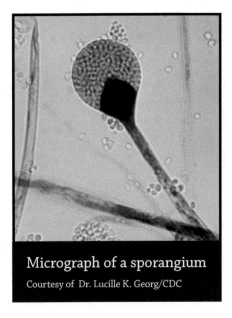
Micrograph of a sporangium
Courtesy of Dr. Lucille K. Georg/CDC

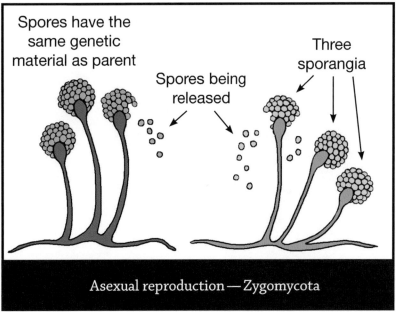
Asexual reproduction—Zygomycota

9.6 Phylum Ascomycota—Yeasts, Truffles

The phylum Ascomycota is named for a reproductive feature called an ascus. During sexual reproduction, a fungus in the phylum Ascomycota produces asci (plural form). Each ascus is a sac that contains microscopic spores called ascospores. When an ascospore is released by the fungus and lands in a location that has the right conditions, a new organism will grow.

Like zygomycetes, the ascomycetes have different mating types. When two hyphae of different mating types touch, they join and form a bridge. The genetic material of one hypha moves into the other hypha. Next, the hypha containing the genetic material from both mating types grows to make more hyphae that will create an ascocarp, the fruiting body of the fungus. When the ascocarp has formed, there will be tips (ends) of hyphae on the ascocarp's inner surface. Asci form on these hyphal tips, and ascospores are produced in the asci. When the ascospores are ready, they will be released from the ascus to grow new organisms that will have genetic material that is different from that of either of the original mating types.

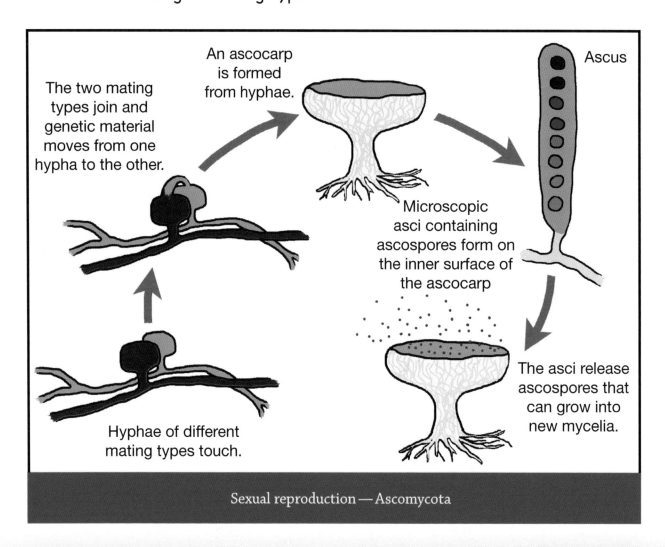

Sexual reproduction — Ascomycota

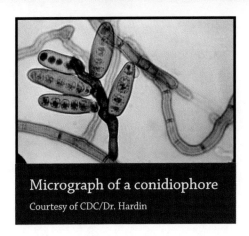

Micrograph of a conidiophore
Courtesy of CDC/Dr. Hardin

Ascomycetes can reproduce asexually by making spores called conidia. The conidia are grouped in structures called conidiophores that grow from the tips of hyphae.

One of the largest of the ascomycetes is the truffle. Truffles are a delicacy in many parts of Europe and North America and are used in soups, stews, and sauces. Truffles can be as small as a pea or as big as an orange. Truffles can be difficult to find because they live as much as 0.3 meter (1 foot) below the surface of the soil. Truffles emit a very strong odor, and truffle hunters are often led to the truffles by trained female pigs which are sensitive to the scent. The truffle that we eat is the ascocarp of the organism.

Truffles

The yeasts we use in food are among the smallest and most familiar members of the phylum Ascomycota. Yeasts are microscopic and reproduce asexually by budding. Under extreme conditions yeasts can reproduce sexually by producing ascospores, although the process is slightly different from that described previously.

Yeasts are used to make many foods such as beer, wine, and bread. When yeast is added to sweetened bread dough, the yeast uses the sugar to make carbon dioxide and ethanol (alcohol). The bubbles of carbon dioxide make the bread dough rise and leave air holes in the baked bread. The alcohol that is being created gives raw bread dough a tangy or sour taste, but the alcohol doesn't stick around for long. When the bread dough is cooked, the ethanol is evaporated by the heat of the oven.

9.7 Phylum Basidiomycota—Mushrooms

The phylum Basidiomycota is named for its reproductive structure which defines the shape of mushrooms. The name Basidiomycota comes from the Latin word *basidium* which means "small pedestal." There are several thousand species of fungi in the phylum Basidiomycota, and these include mushrooms, shelf fungi, puffballs, and rusts.

Like other fungi, basidiomycetes can reproduce asexually or sexually. Asexual reproduction occurs when a hyphal cell goes through the budding process. Recall that during budding, a cell divides in two. The new cell formed by this process separates from the parent cell and has the same genetic material as the parent cell. The new cell can grow into a new organism when the conditions are right. Another type of asexual reproduction in basidiomycetes occurs at the tip of a hypha where a structure (conidiophore) forms to make spores (conidia) that can be released to grow into new organisms.

Basidiomycetes have different mating types that come together during sexual reproduction. When two hyphae that have different mating types touch, they exchange genetic material and make spores that can grow into new mycelia.

In mushrooms, when the underground mycelium is ready to reproduce, a hypha that contains genetic material of different mating types will grow into a knot. Hyphae will then grow from the knot into an above ground mushroom (the fruiting body) that has a stem with a cap, or umbrella, on top. The hyphae that make up the cap grow downward, forming thin ribs called gills. Spores are produced at the ends of hyphae in the gills on the underside of the cap. The spore producing structures are

Mushrooms showing gills on the bottom of the caps

called basidia, which are microscopic and form on the ends of hyphae. The spores made in the basidia are called basidiospores. When they are ready, the basidiospores are released by the mushroom and scattered by the wind. If a spore lands in a spot that has the right conditions, it will grow to make a new mycelium. Because there is such a small chance of a spore landing in a favorable spot, a mushroom will make millions or billions of spores to improve its chances of reproduction.

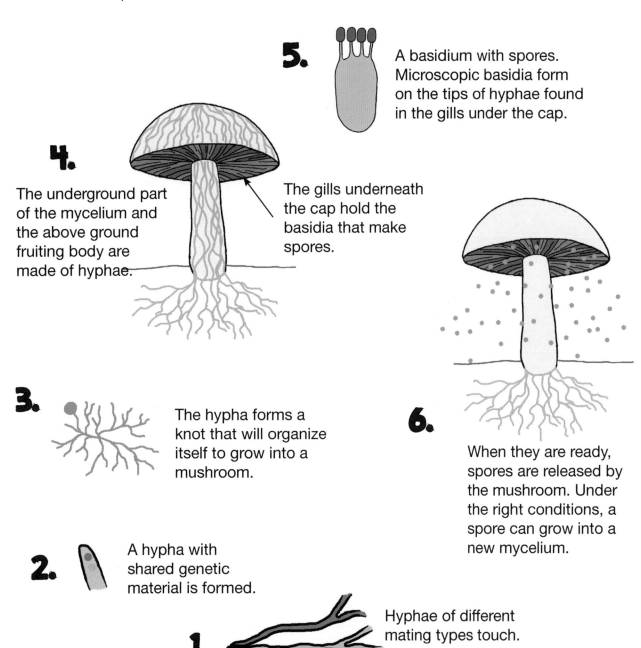

5. A basidium with spores. Microscopic basidia form on the tips of hyphae found in the gills under the cap.

4. The underground part of the mycelium and the above ground fruiting body are made of hyphae.

The gills underneath the cap hold the basidia that make spores.

3. The hypha forms a knot that will organize itself to grow into a mushroom.

6. When they are ready, spores are released by the mushroom. Under the right conditions, a spore can grow into a new mycelium.

2. A hypha with shared genetic material is formed.

1. Hyphae of different mating types touch.

In the right places, you can sometimes find a perfect circle of mushrooms. These circles have the nickname "fairy rings," and the legends say that the fairies danced in a circle there the night before. Can you guess why the mushrooms grow in a circle?

It turns out that a fungal mycelium can live for many years. After a time of growth, as the mycelium gets bigger, the center dies, and the living mycelium becomes a donut shape, sending up mushrooms along its edge. This creates a "fairy ring" of mushrooms.

9.8 Summary

- Fungi are in a kingdom separate from plants, animals, and protists.

- The study of fungi is called mycology.

- With the exception of yeasts, the main part of a fungus is called the mycelium. The mycelium is a mass of tiny, thread-like structures called hyphae that resemble the roots of plants.

- The phylum Zygomycota includes the molds.

- The phylum Ascomycota includes the yeasts.

- The phylum Basidiomycota includes the mushrooms.

Chapter 10 Technology in Physics

10.1	Introduction	96
10.2	Some Basic Physics Tools	97
10.3	Mathematics	101
10.4	Electronics	103
10.5	Computers	106
10.6	CERN	107
10.7	Summary	109

Physics

10.1 Introduction

In a nutshell, physics explores the basic laws of nature. However simple this statement appears to be on the surface, digging deeper we find that physics is a complex aggregate of many different and specialized topics in many areas of science. As a result, physicists require many different types of tools and technology.

The tools physicists need for exploring atoms and subatomic particles are different from those used to study black holes in the farthest reaches of space. Physicists use electronics and computers to study force, speed, energy, gravitation, chaos theory, dark matter, and even game theory. Physicists also use mathematical models, fundamental constants, statistics, standards, and timekeeping to investigate the basic laws of nature. To explore all these areas, physicists have invented a wide variety of different tools and technologies that allow them to study the natural laws that govern how the world works.

In this chapter we will take a small snapshot of just some of the tools and advanced technology used by physicists. We will examine a few core technologies that are used to explore both small and large scale phenomena, and we will also explore a few specialized instruments used for measuring fundamental laws of physics such as force, speed, and energy.

10.2 Some Basic Physics Tools

Measuring Force

Imagine you are pushing a heavy wheelbarrow up a hill. How much force are you using? Do you think it takes more force to squeeze a rubber ball or a marshmallow? How can you tell? Is there a way to measure force?

When you push, pull, squeeze, or elevate an object, you are using force. Recall that force is an action that changes the speed, shape, or position of an object. A force gauge is an instrument that can be used to measure compression, tension, torque, and gravitational force which causes weight. There are several different types of force gauges. Two of these are mechanical force gauges and digital force gauges which are generally handheld instruments that can measure tensile force (pulling) and compression force (pushing). When testing objects, mechanical force gauges and digital force gauges can be held in the hand or mounted on a stand depending on the requirements of the test being performed. The object or objects to be tested are attached to the force gauge which measures the pushing or pulling force being applied to the object.

A mechanical force gauge uses moving parts to measure the amount of force. A needle on a dial indicates how much force is being applied. A digital force gauge uses electronics to measure the amount of force being used and displays the results on a digital screen. Digital force gauges can quickly determine whether the holding strength of a connector or other object is strong enough that it won't break when being used. For example, if you were going to use a rope to lift a heavy box, before you hooked the box to the rope you might want to check how much force the rope can withstand and how much force the box will create on the rope!

Torque gauges measure torque. Torque is the force that is exerted on an object when the object is being rotated around an axis. Torque gauges measure how much torque is being created. For example, when you use a wrench to attach pedals to your bike, you are applying torque with the wrench. A torque wrench, a wrench that contains a torque gauge, is a great way to determine the point at which you are applying just enough torque but not too much!

Measuring Speed

How fast can you run? How fast can a car go? How fast does a bullet go? How fast does an electron travel? How can you measure the speed of a runner, a car, a bullet, or an electron? As we will see in Chapter 12, speed is the distance traveled divided by time, so both distance and time need to be measured to determine speed.

Speedometer

Odometer

In automobiles a speedometer is used to measure speed. A speedometer is a type of gauge that works from the car's driveshaft—the shaft used to turn the wheels. As the driveshaft turns, the speedometer measures how fast the driveshaft is rotating the wheels and uses this information to calculate the number of miles per hour being traveled. The odometer is the gauge that displays the distance traveled which can be calculated by multiplying the number of wheel rotations by the size of the tire.

Physics—Chapter 10: Technology in Physics 99

If you know the distance traveled, speed can be calculated by using a stopwatch. A stopwatch measures time and can be started and stopped at different intervals. Stopwatches can have a digital screen that displays a number or a dial with a needle that moves. Stopwatches can be used for measuring speed during a physics experiment and on the track field to tell how fast sprinters and distance runners are going.

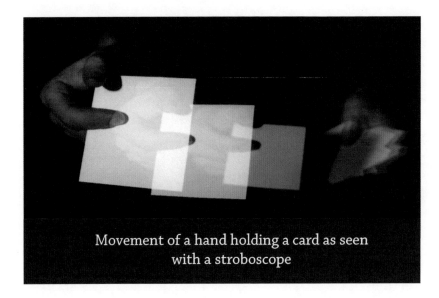
Movement of a hand holding a card as seen with a stroboscope

A strobe light or stroboscope is used to measure the speed of a rotating motor, fan, or other object. A strobe light flashes on and off very rapidly causing the movement of the object to appear as a series of snapshots rather than a continuous flow. This can produce the illusion of the object moving in slow motion. If a particular place on a motor or fan is marked with a dot, a strobe light or stroboscope can be used to help measure how fast the object is turning. The speed of rotation can then be adjusted to the desired speed.

Lasers are tools that emit a narrow beam of concentrated light and can be used to measure speed. If you've ever been in a car that was going over the speed limit and passed a parked police car only to find the police car soon following you with warning lights blazing, you just had your speed measured with a laser speed gun. A laser speed gun works by bouncing light off a moving object.

100 Exploring the Building Blocks of Science: Book 6

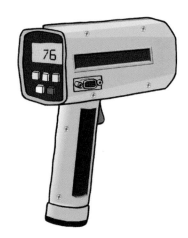

Light is emitted from the speed gun in a narrow, concentrated beam. The light reaches the moving object, bounces back, and is detected by the speed gun. The speed gun calculates the speed of the moving object by using the time required for the light to bounce back to it.

Measuring Energy

Tools for measuring energy are very important in physics. Recall that energy can come in a variety of forms including heat energy, electrical energy, chemical energy, nuclear energy, mechanical energy, and light energy.

Thermometers are used to measure temperature. For many materials, heat energy is closely related to temperature. A bulb thermometer uses the expansion of materials such as mercury or alcohol to measure heat energy by displaying the amount of heat energy as temperature. Both alcohol and mercury will expand (become bigger) when heated and contract (become smaller) when cooled. By placing alcohol or mercury in a thin tube that has graduated markings, the temperature (which is related to heat energy) can be measured.

Electronic thermometers can also be used to measure temperature. An electronic thermometer contains a battery that creates an electric current, and it also contains a thermal sensor (heat sensor) called a thermistor. A thermistor is a component made of a special ceramic or polymer material that senses heat. This material is also a resistor that allows more or less electricity to flow through the thermistor according to the temperature being measured. A simple computer in the thermometer translates the amount of electrical flow into a temperature and displays this number on a screen. Electronic thermometers are used to measure the heat of an oven, the temperature of a car engine, and even the temperature of living things like animals and humans.

Voltmeters are used to measure electrical energy. A voltmeter measures the voltage difference between two points in an electric circuit. Voltage is a measure of electromotive force (EMF). Voltmeters have a needle gauge or digital display that shows the voltage difference between its two leads. Portable voltmeters have leads that can be attached to positive and negative terminals. For example, you can use a voltmeter to measure the voltage (potential electrical energy) inside a battery by placing one lead on the positive battery terminal and one on the negative terminal.

Voltmeters can often be switched between measuring voltage and measuring current. Current is the rate of flow of electrical charge between two points. Current and voltage are related but are not the same thing. We will learn more about voltage and current in Book 7 of this series.

10.3 Mathematics

One of the most amazing and beautiful features of physics is that physical laws can be expressed precisely using mathematics. The use of mathematics is an essential tool in physics. Without mathematical calculations we would not be able to see distant stars, explore the depths of the oceans, land a probe on a comet, or visualize atoms and molecules.

One of the most basic mathematical tools used in physics is geometry. Geometry uses distances and angles to describe the exact relationship of objects to each other. The mathematics behind geometry is largely attributed to the Greek mathematician Euclid (circa 325 BCE-circa 265 BCE). Euclid's geometry, called Euclidean geometry, is based on a set of axioms. An axiom is a statement that is used as a starting point, premise, or postulate. For example, one of Euclid's axioms states that "any line segment can be extended indefinitely in a straight line." As we will see in Chapter 13 on non-linear motion, we can use this axiom to calculate the velocity of a stone that has been thrown into the air.

EUCLID
Circa 325-265 BCE

Algebra is also an important mathematical tool for physics. Algebra is a type of mathematics that uses symbols in arithmetic calculations. Some historians think that algebra was likely developed by the Persian mathematician al-Khwarazmi (circa 780-circa 850 CE) and then further developed by the French mathematician Francois Viete (1540-1603 CE) in the late 16th century.

Algebra is used to relate and solve a variety of relationships in physics. For example, the speed of an object is defined as:

$$\text{speed} = \frac{\text{distance}}{\text{time}}$$

If you want to calculate the speed of a car going from San Francisco to Los Angeles, you can plug in the values for "distance" and "time."

$$\text{speed} = \frac{382 \text{ miles}}{6 \text{ hours}}$$
$$\text{speed} = 64 \text{ miles per hour}$$

But what if you know you can only go 50 miles per hour? How can you find out how long it will take? With algebra you can generalize the equation using mathematical symbols. Then you can rearrange the equation to solve for the number of hours.

Using mathematical symbols the equation now becomes:

$$s = \frac{d}{t}$$

where "s" represents "speed", "d" represents "distance" and "t" represents "time." This is a simple algebraic equation.

Applying the rules of algebra, we can rearrange the symbols and solve for time "t":

$$t = \frac{d}{s}$$

Plugging in the values for distance (d) and speed (s) you have:

$$t = \frac{382 \text{ miles}}{50 \text{ miles per hour}}$$
$$t = 7.64 \text{ hours}$$

More complex physics requires more complex mathematical tools. Trigonometry, differential equations, complex algebra, and vector calculus are all complex mathematical tools used in physics. Without the use of mathematics, physics would still be considered a "philosophy" rather than a science in which exact quantities can be measured, evaluated, and calculated.

10.4 Electronics

Discoveries about how electrical currents work have made many advancements in science possible. The development of electronic equipment for physics is a great example of science influencing technology and technology influencing science.

Electricity is the set of phenomena that arises from electrical charges and movement of electrical charge. Electric charge in the form of electrons can move through metals, semi-conductor materials, and even a vacuum and can produce electrical power, electric potential, and electromagnetic fields.

The nature of moving and static electric charge was an area of active investigation for many early scholars. People in ancient times knew that certain objects could produce an electric shock, but they weren't able to determine why this happened. It wasn't until the late 17th century that theories about atoms were developed and explored. Although experiments were performed to investigate electrical currents, the connection between electrons and atoms wasn't discovered until much later.

Vacuum tube; Integrated circuit (Courtesy of CSIRO)
Two transistors (Photos courtesy of Charles Rondeau)

As information about atoms, electrons, protons, and neutrons became available, understanding of electrical phenomena increased. Experiments using electrical currents led to the development of complex technological components including the vacuum tube, transistor, and integrated circuit. As this technology continued to be developed, more physics theories could be explored. The cycle between technology advancing science and science advancing technology continues today.

A vacuum tube is a glass tube or bulb that controls electric current through a vacuum that is sealed inside the tube. The invention of the vacuum tube made it possible to manipulate, amplify, and transmit electrical energy. The vacuum tube was used in early radio communications and made development of radio possible. Vacuum tubes were also used in early computers but were replaced when the transistor was invented. Transistors function like the vacuum tube, manipulating, amplifying, and transmitting electrical energy, but transistors are small, weigh less, and require less power to run. Integrated circuits regulate electrical energy by combining transistors and other electrical components such as capacitors that store electric charge and resistors that limit the flow of electric current. Equipment is called electronic when it contains these kinds of small electrical components.

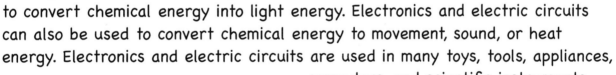

Electronic instruments use the movement of electric charge through an electrical circuit to create a variety of different outcomes. Often electronics are used to convert energy from one form to another. For example, a flashlight uses electronic components to convert chemical energy into light energy. Electronics and electric circuits can also be used to convert chemical energy to movement, sound, or heat energy. Electronics and electric circuits are used in many toys, tools, appliances, computers, and scientific instruments.

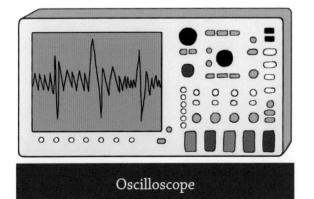
Oscilloscope

Electronics found in many modern physics labs include simple instruments, such as stopwatches that measure time, to more advanced instruments, such as those that detect the number of atoms on a silicon surface. Physicists use electronic oscilloscopes to observe how voltage signals vary with time, voltmeters to measure the amount of voltage, and spectrometers to measure light. The development of electronics has made all of these instruments possible.

10.5 Computers

Computers play a vital role in physics labs today. Most electronic equipment can be attached to a computer. Computers allow physicists to quickly collect, modify, and analyze data. Computers also allow physicists to create computer simulations and models of physical data. These simulations and models help scientists understand and predict a variety of physical phenomena.

For example, physicists and engineers can use computer models to explore airplane designs and can use a flight simulator to find out how an airplane might react during extreme weather.

A robot can work under dangerous conditions

Courtesy of DARPA

Computers and computer models are also used in the field of robotics. Exploring how to convert electrical energy into mechanical energy is important in robotics. Having a robot process information as easily as a human is one of the many goals of robotics. With computers and computer models physicists can determine how to create robotic movements that copy human legs for walking and the human hand for grasping objects.

Computers are used in optics, laser, and microscopy labs. The fine adjustments needed to control a scanning tunneling microscope (STM) are generated by a computer that sends electrical signals to a piezoelectric crystal that moves a tiny sample by very small amounts. Because such small movements are possible, individual atoms can be seen.

10.6 CERN

By putting new theories and discoveries in physics together and using the most advanced technology and mathematics, scientists can now explore the very fundamental structure of the universe. At the laboratory that straddles the border of France and Switzerland, the European Council for Nuclear Research (CERN) has assembled a technological masterpiece for probing the basic constituents of matter and the forces that act on them.

Aerial view of CERN - circle shows the location of the underground Large Hadron Collider

Courtesy of CERN/ Maximilien Brice, photographer

At CERN, whose name comes from the acronym for the French "Conseil Européen pour la Recherche Nucléaire," physicists are exploring theories about the universe, such as antiprotons (protons with a negative charge), dark matter (invisible, as yet unknown matter that makes up most of the universe), particles called quarks and leptons that are thought to be smaller than protons and neutrons, and a very special particle called the Higgs boson that physicists think may be what gives matter its mass.

Thousands of scientists and engineers from many countries all over the world have worked with CERN to build and operate a huge instrument called the Large Hadron Collider (LHC) that explores what happens when parts of atoms (particles) are smashed together at very high speeds. The Large Hadron Collider is a particle accelerator in which beams of particles shoot through the collider's tunnel—a structure that is underground, donut shaped, and 27 kilometers (17 miles) in circumference.

The Large Hadron Collider contains two very sophisticated pieces of equipment—an accelerator and a detector. The accelerator creates two beams of protons from bottled hydrogen gas and pushes the protons to go very fast—so fast that they are traveling at nearly the speed of light, making 11,000 circuits per second around the tunnel! The two beams are traveling in opposite

Inside the Large Hadron Collider tunnel
Courtesy of CERN/ Maximilien Brice, photographer

directions, and once the protons are moving at nearly the speed of light, the two beams of protons are made to collide with each other. When particles smash into each other, a detector gathers up all the information. There are four detectors that determine if the protons have smashed into other particles, and if so, how fast they were going, how big they were, and their charge. Data from the detectors is sent to the CERN Data Centre and from there it is sent through the Worldwide LHC Computing Grid to labs in many different countries, giving thousands of physicists access to the data.

Physicists can use the Large Hadron Collider to study a specialized field of physics called particle physics. Using data from the LHC, physicists can study how cosmic rays affect Earth's atmosphere and cloud formation, how antiprotons interact with living things, and how particles are generated by the Sun.

10.7 Summary

- Physics includes many different and specialized topics and requires many different types of tools and technology.

- Some basic physics tools allow physicists to study force, speed, and energy.

- The use of mathematics is an essential tool in physics, and physical laws can be described exactly through mathematics.

- The study of physics uses advanced, specialized electronic equipment along with simpler instruments.

- Electronics and computers play a vital role in physics.

- CERN's Large Hadron Collider is helping physicists explore the fundamental structure of the universe.

Chapter 11 Motion

11.1	Introduction	111
11.2	Inertia	112
11.3	Mass	112
11.4	Friction	113
11.5	Momentum	114
11.6	Summary	116

11.1 Introduction

What is motion? We see objects moving every day. Cars move down the road. Planes move in the sky. We move as we work and play. Moving seems to be a very ordinary thing. However, figuring out exactly how objects move was a problem for early scientists. Although many people tried to figure it out, it took nearly 2000 years to finally understand!

According to history, the first person to study the science of motion was Aristotle. Aristotle was a Greek philosopher who was born in Stagira in 384 BCE. Aristotle thought that every moving thing was being constantly pushed from behind. He thought that the air in front of a moving ball was being separated as the ball moved and that the air behind would close up forcing the ball forward. He also thought that all moving objects moved because of this constant force. Based on these ideas, he thought that since he didn't feel the Earth moving, the Earth was sitting still. He also thought that the Sun and stars, because they changed places in the sky, were moving around the Earth. This belief system is based on a geocentric cosmos. *Geo* is the Greek word for "earth," and *centric* means "central." So geocentric means "earth centered." *Cosmos* is the Greek word for universe, so a geocentric cosmos means "earth-centered universe."

Today we know that the Earth moves around the Sun and that our solar system is heliocentric or "sun-centered." [*helios* is the Greek word for "sun"].

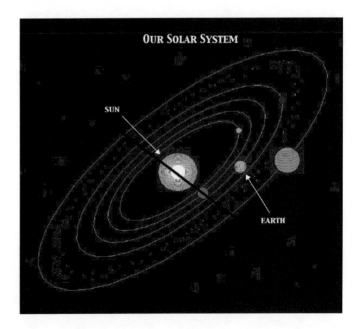

Although Aristotle had many good ideas, he was wrong about motion. In the early 1600s Galileo finally showed with experiments that motions do not require a constant force to keep them going, and later Isaac Newton put these concepts into mathematical terms.

11.2 Inertia

If there is nothing pushing on a ball as it travels through the air, as Aristotle thought, how does it keep going? What keeps the ball moving at all? What Galileo discovered was that things always keep moving unless something stops them. This property is called inertia.

Simply put:

Inertia is the tendency of things to resist a change in motion.

This means that once something is moving, it will not stop, slow down, or change its direction unless something pushes on it. Aristotle had it completely backwards! It's not that forces keep things moving, but that forces *stop or change* the speed or direction of things that *are* moving!

Everything has inertia, no matter what it is—an atom, a rock, a baseball, or a car. Once the object gets going, it won't stop or even change direction unless something pushes on it.

11.3 Mass

The property that gives things inertia is called mass. Everything has mass, so everything has inertia. Because the force of gravity is constant everywhere on Earth, you can tell how much mass something has by weighing it.

For example, you can tell that a marble has less mass than a baseball because a marble weighs less than a baseball, and you can tell that a baseball has less mass than a bowling ball because a baseball weighs less than a bowling ball.

We know that once rolling, a marble would be much easier to stop than a baseball, and a baseball would be much easier to stop than a bowling ball. Can you imagine trying to bowl with marbles or trying to play marbles with bowling balls? The marbles don't have enough mass to knock down the pins, and the bowling balls have too much mass to roll with your thumb!

11.4 Friction

So, if inertia keeps things moving, what makes things stop? If you roll a ball, or push a toy truck, or take your foot off the gas pedal in your car, the ball, the truck, and the car all eventually stop moving. Why? If Galileo was right and inertia keeps things going, then wouldn't everything keep going all the time? Why do things stop?

What Galileo discovered is that objects keep going if no forces act on them. However, everyday objects like cars, bowling balls, and toy trucks almost always have forces acting on them. The force that makes objects stop is called friction. Friction is a force that tends to slow things down.

Usually friction is caused by things rubbing against each other. For example, if you slide a hockey puck on the street, the atoms in the hockey puck rub against the atoms in the street. The rubbing of these atoms against each other causes frictional force. The rougher the two surfaces, the more friction there is. If you could change the way the hockey puck contacts the street, you could reduce the friction. This is why hockey is not usually played on streets, but on ice. The ice is much smoother than the street, and the hockey puck slides much more easily. Galileo discovered that if he got rid of friction, things would keep on going and never stop. For example, if you could play hockey in space where there is no air and no friction, the puck would never stop! This is one of the most important discoveries in all of physics.

11.5 Momentum

Things are also harder to stop if they move fast. A baseball rolling slowly on the ground is easy to stop, but a baseball thrown by a pitcher is much harder to stop and requires a padded glove. So, there are two things that make something hard to stop: mass and speed.

In physics, the property that makes things hard to stop is called momentum. It has a precise mathematical definition:

$$\text{momentum} = \text{mass} * (\text{speed} + \text{direction})$$
$$\mathbf{p} = m*\mathbf{v}$$

Where "**p**" stands for momentum, "**m**" for mass, and "**v**" for velocity. Velocity is speed + direction. We will learn more about velocity in the next chapter. (Note: "**p**" is the symbol for momentum because "**m**" is already being used for mass. "**p**" comes from the Latin word *petere* which means "to pursue or go towards.")

You can see from this equation that the more mass something has, the more momentum it will have. Also, the faster something travels (the more speed it has), the more momentum it will have. If something has lots of momentum, it will be harder to stop than something that has little momentum.

Momentum can change whenever there is a change in the mass of an object, the velocity, or both. But there is another factor that is important for changing momentum and that is time. Remember from Section 11.2 that forces stop or change the speed or direction of things that are already moving. If you apply a force for a longer period of time, you create a bigger change in speed or direction and thus a greater change in momentum.

For example, imagine that you are in a Soap Box Derby and your job is to get your team's go-kart moving fast enough that you can win. If you give the go-kart one strong push, it's not going to go very far. However, if you push the go-kart for a few yards with the same amount of force, the go-kart will travel farther. Why? The go-kart travels farther because you are changing velocity over a longer period of time, resulting in more momentum.

Knowing how time affects momentum can be useful. Understanding that momentum can increase as the length of time increases is great if you want to get a go-kart moving faster. But knowing that momentum can decrease as the length of time decreases can be handy in certain situations.

Momentum Math!

The relationship between **velocity**, **force**, and **time** can be described using mathematics and Newton's Laws. Newton figured out the following equation to show that **force**, **mass**, and a **change in velocity** over time were related:

Force = (mass) * (change in velocity ÷ time)

If we use some mathematics to rearrange this equation by multiplying both sides by time we get:

Force * time = change in (mass * velocity)

Because a change in **mass** * **velocity** equals **momentum** we get:

Force * time = momentum

This shows that the longer the time that a force is applied the greater the momentum will be.

Imagine you are sledding down a long hill and you discover you are going too fast to stop using only your feet. You can steer your sled into a fluffy pile of snow to the left or the side of a barn to the right. Which would you choose? You will hopefully choose the fluffy pile of snow because you know that the fluffy pile of snow will slow your momentum over a greater amount of time than the side of a barn, and by slowing your momentum more gradually you create less impact (force) as you stop, which is much easier on the body!

11.6 Summary

- Inertia is the tendency of objects to resist change in motion.

- Friction is a force that tends to slow objects down.

- Momentum depends on both the mass of an object and its velocity (speed + direction).

- Objects with large mass have more momentum than objects with small mass.

- Objects with a lot of speed have more momentum than objects with little speed.

- For a given applied force, momentum increases as length of time increases and decreases as time decreases.

Chapter 12 Linear Motion

12.1	Introduction	118
12.2	Speed	118
12.3	Velocity	124
12.4	Acceleration	128
12.5	A Note About Math	130
12.6	Summary	131

Physics

12.1 Introduction

How fast can you ride your bike? Can you race a car on your bike? Can you race a train? How fast does an airplane go? Does it go faster than a car? Does it go faster than a rocket? How can you measure how fast a bike, a car, or an airplane goes?

In the last chapter we looked at some general features of motion. We explored inertia and how an object will stay still or stay in motion because of inertia; how mass, momentum, and speed are related; and how friction will change or slow down the motion of an object. In this chapter we will take a closer look at a particular type of motion called linear motion.

Linear motion is, very simply, motion that occurs when any object travels in a straight line. In this chapter we will see that mathematics can be used to describe motion and that motion is defined by speed, acceleration, and velocity.

12.2 Speed

If you were to hop on your bike and cycle to the nearest grocery store for chocolate milk, how long would it take you? How long would it take to get to the same grocery store by car? Maybe you live in the remote wilderness and your only way to get to the store is by plane. How long would it take you to travel to the store by plane?

Whether you travel by bike, car, or plane, in each case you are going from one point to another. In other words, you are traveling a certain distance. If the distance is short enough, you could ride your bike. However, if the distance is far, you would probably want to take a car or even a plane.

The reason to choose a car or plane instead of a bike to travel a far distance is that a car or plane can go faster than a bike. In other words, a car or plane travels at a higher speed than a bike, and therefore a car or plane can travel a longer distance in less time than a bike.

Using mathematics we can calculate exactly how fast a car, plane, or bike can travel a certain distance. Speed is defined as the rate at which an object covers a given distance in a given amount of time. Recall from Chapter 10 that speed is defined mathematically by the following equation:

$$s = \frac{d}{t}$$

where "s" represents "speed," "d" represents "distance," and "t" represents "time."

To see how this equation allows us to calculate speed, let's imagine that you want to bicycle to the grocery store which is 16 km (10 miles) away. Because you're in really good shape, it only takes you 30 minutes to get to the store to buy chocolate milk. How fast did you go?

If we plug in the value of 16 km (10 miles) for distance and 0.5 hours for 30 minutes we get:

$$s = \frac{16 \text{ km (10 miles)}}{0.5 \text{ hrs}} = 32 \text{ km per hour (20 miles per hour)}$$

Suppose you don't know how long it might take you to ride your bike to the grocery store. But you do know how far it is to the store, and you know about how fast you can ride your bike. Can you figure out how long it will take? It's easy! All you have to do is rearrange the equation and *solve for time*.

Some Basic Math Rules, Terms, & Symbols

These symbols all mean divide: —, /, ÷

These formulas all mean that speed equals distance divided by time:

$$s = \frac{d}{t} \qquad s = d/t \qquad s = d \div t$$

These symbols all mean multiply: x, *, ·

These formulas all mean that momentum equals mass times velocity:

$$\mathbf{p} = \mathbf{m} \times \mathbf{v}, \qquad \mathbf{p} = \mathbf{m} * \mathbf{v}, \qquad \mathbf{p} = \mathbf{m} \cdot \mathbf{v}$$

v   bold lower case "v" represents velocity
p   bold lower case "p" represents momentum
Δ   "delta" symbol—represents the difference between two quantities
| | "absolute value"—these upright lines mean that the result of a formula contained within them must be expressed as a positive number

variable - In an algebraic formula, a variable is a letter that represents a number that can change (vary).

1. In algebra when we perform the same operation to both sides of an equation, the two sides of the equation remain equal. We can add, subtract, multiply, or divide the same number (or variable) to both sides of the equation and the equation stays equal.

    For example, because $s = \frac{d}{t}$, then $s*2 = \frac{d}{t}*2$, or $s*x = \frac{d}{t}*x$, or $s*t = \frac{d}{t}*t$

2. In algebra if we multiply a variable by its inverse, we get 1.

    The inverse of a variable is its opposite, so the inverse of:

    $\frac{s}{1}$ is $\frac{1}{s}$ and $\frac{t}{1}$ is $\frac{1}{t}$ and $\frac{d}{1}$ is $\frac{1}{d}$

    If we multiply a variable by its inverse we get 1. For example:

    $\frac{s}{1} * \frac{1}{s} = \frac{s}{s} = 1 \qquad \frac{t}{1} * \frac{1}{t} = \frac{t}{t} = 1 \qquad \frac{d}{1} * \frac{1}{d} = \frac{d}{d} = 1$

    Any number divided by itself equals 1.

3. We can rearrange either side of an equation to help solve it. For example:

    $s*t = t*\frac{d}{t}$ can be rearranged to $s*t = \frac{t}{t}*d$

    Because... $\frac{t}{t} = 1$ we get... $s*t = 1*d$ so... $s*t = d$

Let's imagine that you want to go to the store in the next town because they have the most delicious chocolate milk. However, this store is 64 km (40 miles) away. If you can ride your bike at 32 km (20 miles) per hour, how long will it take to get to the store?

Here's how we can use some basic math (algebra) to *solve for time*. First we multiply both sides of the equation by "t" (time):

$$s = \frac{d}{t} \quad \text{(speed = distance divided by time)}$$

$$s*t = t*\frac{d}{t} \quad \text{(both sides multiplied by time)}$$

Because $t * \frac{1}{t} = 1$, the "t"s on the right hand side cancel each other out:

$$s*t = \cancel{t}*\frac{d}{\cancel{t}}$$

Results: $\quad s*t = d \quad$ (speed multiplied by time = distance)

Now, if we divide both sides by "s" (speed) we get:

$$\frac{s*t}{s} = \frac{d}{s}$$

Because $s * \frac{1}{s} = 1$, the "s"s on the left hand side cancel each other:

$$\frac{\cancel{s}*t}{\cancel{s}} = \frac{d}{s}$$

Results: $\quad t = \frac{d}{s} \quad$ (time = distance divided by speed)

Using this equation we can figure out how long it will take to go 64 km (40 mi.) at 32 km (20 mi.) per hour. Plugging in the values for "d" and "s" we get:

$$t = \frac{64 \text{ km}}{32 \text{ km per hour}} \quad \text{(or} \quad t = \frac{40 \text{ miles}}{20 \text{ miles per hour}} \text{)}$$

Solution: $\quad t = 2$ hours

Because the trip will take two hours, you might want to grab an extra chocolate milk for the ride home! Or, if you don't have four hours for a trip to the grocery store and back, you could go by car.

You Do it!

1. Your car will go 80 km per hour (50 miles per hour). How long will it take you to get to the grocery store and back if the store is 80 km (50 miles) away?

2. Which bicycle route will get you there faster?
   Route A: 30 km with an average speed of 15 km per hour,
   or
   Route B: 30 km with an average speed of 10 km per hour

See solution steps below.

Solution Steps

1. **Calculate for time**

   80 km/h (50 mph) = speed = s
   80 km (50 miles) = distance = d
   We are looking for "time" = t

   To solve:

   $$speed = \frac{distance}{time} \qquad s = \frac{d}{t}$$

   Multiply both sides of the equation by "t":

   $$s * t = \frac{d * t}{t}$$

   The "t"s cancel:

   $$s * t = \frac{d * \cancel{t}}{\cancel{t}} \qquad s * t = d$$

   Divide both sides by "s":

   $$\frac{\cancel{s} * t}{\cancel{s}} = \frac{d}{s} \qquad \boxed{t = \frac{d}{s}}$$

   The "s"s cancelled so this is my equation!

   Plug in variables:

   $$t = \frac{80 \text{ km}}{80 \text{ km/h}} = 1 \text{ hour}$$

2. **Route A or B?** (See next page for solution steps.)

> ## Solution Steps (continued)
>
> 2. Route A or B?
>
> I'll use the equation for time.
>
> **To solve for Route A:**
>
> $d = 30 \text{ km}$
>
> $s = 15 \text{ km/h}$
>
> $t = \dfrac{d}{s} = \dfrac{30 \text{ km}}{15 \text{ km/h}} = \boxed{2 \text{ hr}}$ *Route A is faster!*
>
> **To solve for Route B:**
>
> $d = 30 \text{ km}$
>
> $s = 10 \text{ km/h}$
>
> $t = \dfrac{d}{s} = \dfrac{30 \text{ km}}{10 \text{ km/h}} = 3 \text{ hr}$

In the previous example, we assumed a **constant speed**. In other words, we assumed that there are no stop signs or slow traffic or any other reasons to go faster or slower on our trip to the grocery store.

However, on most roadways there are stop signs and traffic congestion that will change the speed you are traveling. If you are riding a bicycle, there may be hills that cause you to go slower on the way up and faster on the way down.

When we use the total distance traveled divided by the total time it takes, we are actually calculating the **average speed**. The speed of the car or bicycle will change depending on whether or not there are other cars or bicycles on the road, turns to make, and traffic signs to obey. By using the total time and total distance, we don't see the changes in speed from one section of the trip to the next. However, if you have a speedometer, you can watch how fast or slow you go in certain sections of the trip and calculate how much time it would take you to travel only those sections. Knowing how long it might take to travel certain sections might change the route you take.

For example, suppose your house is in the middle of the block and in the middle of a steep hill. To get to the store you first have to get out of your neighborhood, and you can choose to either go to the north and up the hill or to the south and down the hill. Even if the distance is longer, you might decide to go south, riding a little farther but going downhill so you can ride faster. Your average speed would be different for each route because the speed you travel over a particular section would be different.

12.3 Velocity

In the last example we added "north" and "south" to our discussion of speed. North, south, east, and west are all directions. A direction describes where you are headed relative to your current position. If you say, "I am going north to the bowling alley," you are describing the direction of the bowling alley relative to your current position.

In physics, speed plus direction is called velocity. If you say you are going 25 km per hour, you are describing your speed. However, if you say you are going 25 km per hour to the north, you are describing velocity.

Although speed and velocity look similar, in physics they are actually quite different. Both speed and velocity have distinct but different mathematical meanings. Speed is a scalar quantity and velocity is a vector quantity. (See next page for definitions.)

Scalars and Vectors

Mathematically, speed is called a scalar quantity. In math, scalar simply describes an amount or magnitude. For example, if a car is traveling at 40 km per hour, this describes the *amount* of speed at which the car is traveling from one point to another. A scalar is simply a number that represents a value like speed, temperature, weight, or height. In an equation, a scalar is written as a lower case letter. For example: s, d, t

Mathematically, velocity is called a vector quantity. In math, a vector is both the magnitude (amount) and the direction (up, down, right, left, north, south, etc.). A vector has spatial dimensions (in 2, 3, or more dimensions), a direction, and a magnitude. In an equation, a vector is written as **v** (a lower case letter in a **bold** font).

We can draw a vector as an arrow. The length of the arrow illustrates the magnitude, and the orientation of the arrow illustrates its direction. Drawing an arrow on a graph will also illustrate its spatial location.

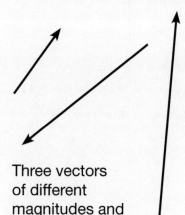

Three vectors of different magnitudes and orientations

Speed (a scalar) is the magnitude (numerical value) for velocity (a vector). The length of the arrow represents the numerical value of the magnitude.

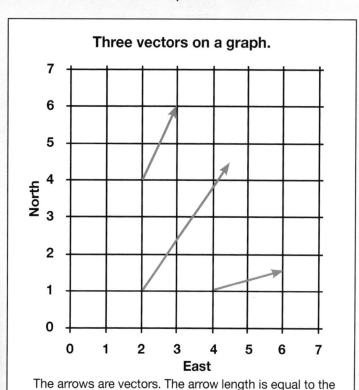

Three vectors on a graph.

The arrows are vectors. The arrow length is equal to the vector's magnitude, and the direction of the arrow shows the vector's orientation.

Technically speaking, speed is the rate at which an object moves, and velocity is the rate at which an object changes its position.

On a linear path the speed and velocity have the same magnitude (or amount). If a horse going north trots 13 km (8 miles) in one hour from point A to point B, the horse's speed is:

$$s = \frac{d}{t} = \frac{13 \text{ km}}{1 \text{ hr}} = 13 \text{ km per hour}$$

And the velocity is:

$$v = \frac{d}{t} = \frac{13 \text{ km}}{1 \text{ hr}} = 13 \text{ km per hour north}$$

What happens if a horse is trotting on a circular path? Because velocity takes direction into account, speed and velocity are different.

Imagine the horse is on a 1 km (.6 mile) track and trots around the track 13 times, traveling 13 km in one hour and returning to the starting line. The speed, or rate, at which the horse trotted is:

$$s = \frac{d}{t} = \frac{13 \text{ km}}{1 \text{ hr}} = 13 \text{ km per hour}$$

However, since the horse stopped and started at the same place, there was no change in its position and the velocity would be zero! (See the math on the next page.)

Strange but True Velocity Math!

In the example of the horse trotting around the track, we can use an illustration to see how the velocity is equal to zero. If we divide the track up into different sections and add the velocities of each section, it looks like this:

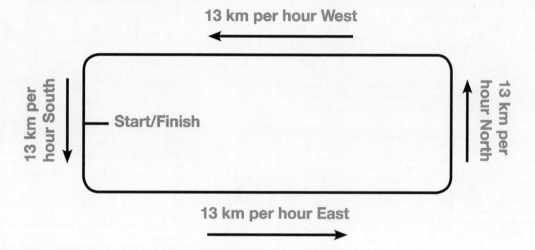

The total velocity of the horse is:

Total v_T = 13 km/h north - 13 km/h south + 13 km/h east - 13 km/h west

$$v_T = v_n - v_s + v_e - v_w$$

Because velocity is a vector, the directions cancel each other and we see that:

$$v = 0$$

In this example the velocity of the horse is zero even though it maintained a constant speed of 13 km per hour. When the horse returned to its starting position, there was no change in its position from when it began trotting, so the total velocity is zero.

12.4 Acceleration

Sometimes velocity changes. If you are riding your bike and come to a big hill, you might slow down as you climb up. When you go over the top and then have a long descent, you will most likely go faster. In both cases your velocity will change because both your direction and speed change as you slow down or speed up and go up or down the hill.

You have probably felt the effects of going faster or slowing down on a bike, in a car, or in a plane. If you are racing to the finish line on your bike, you might spin your legs faster to win! When you spin faster, you can feel your body jerk a little as you suddenly speed up. If you are in a car and the light turns yellow, you might press on the brake to slow down to a full stop. As you quickly push on the brake, you can feel your body move forward. In both cases you can feel the moment when you suddenly change velocity.

We change the velocity of something by changing its speed, its direction, or both its speed and direction. When velocity changes over a given time it is called acceleration. Acceleration can be represented by the equation:

$$a = \frac{\Delta v}{\Delta t}$$

where "a" represents acceleration, "Δ (delta) v" represents the change in velocity and "Δ (delta) t" represents the change in time. Acceleration equals the change in velocity divided by the change in time.

The small delta symbol (Δ) is a mathematical symbol that represents the difference between two quantities. In this equation "Δv" stands for the difference between the final velocity and the initial velocity, and "Δt" stands for the difference between the final time and the initial time. When we expand the equation by plugging in the variables for the initial and final velocities and times, we get:

$$a = \frac{v_f - v_i}{|t_f - t_i|}$$

Where "v_f" is the final velocity, "v_i" is the initial velocity, "t_f" is the final time, and "t_i" is the initial time. Also, for acceleration, time is always a positive number. Δt represents the "change in time." When the result of a calculation must be a positive number, or absolute value, the calculation is placed between two upright lines. In the acceleration formula, the change in time is written as $|t_f - t_i|$ to show that the result is expressed as a positive number.

Acceleration MATH!

Example 1:

Calculate the acceleration of a bowling ball going from zero to 2 meters per second in 2 seconds.

$$a = \frac{v_f - v_i}{|t_f - t_i|} = \frac{2 \text{ m/sec.} - 0 \text{ m/sec.}}{|2 \text{ sec.} - 0 \text{ sec.}|} = \frac{2 \text{ m/sec.}}{2 \text{ sec.}} = 1 \text{ m/second}^2$$

Example 2:

Calculate the acceleration of a projectile slowing down from 50 meters per second to 10 meters per second in 10 seconds.

$$a = \frac{v_f - v_i}{|t_f - t_i|} = \frac{10 \text{ m/sec.} - 50 \text{ m/sec.}}{|10 \text{ sec.} - 0 \text{ sec.}|} = \frac{-40 \text{ m/sec.}}{10 \text{ sec.}} = -4 \text{ m/second}^2$$

You Do the Acceleration MATH!

Physics Math Problem 1

1. Calculate acceleration where $v_f = 10$ m/sec., $v_i = 25$ m/sec., $t_f = 5$ seconds, $t_i = 0$
2. Calculate acceleration where $v_f = 40$ m/sec., $v_i = 10$ m/sec., $t_f = 10$ seconds, $t_i = 0$

(See Appendix for solution.)

12.5 A Note About Math

You can see in this chapter that you can describe linear motion exactly with mathematics. Learning mathematical terms like scalar and vector, symbols like delta, and algebra rules can make linear motion both clearer and more complicated, depending on how much math you understand.

As we discussed in Chapter 10, mathematics is an essential tool for physics. Because physical actions like linear motion can be described exactly using math, scientists have been able to launch rockets into space, put people on the Moon, and explore the possibility of traveling to other worlds.

The best way to learn the math is to practice. By understanding and solving the problems in this book and making up your own problems over and over again, you can master physics and math!

Physics—Chapter 12: Linear Motion

More Math!

Physics Math Problem 2

Imagine you are riding your bike in the Tour de France and you come to the famous mountain, the Col du Galibier. It is full of hairpin turns and steep inclines as the road winds up to Plan Lachet.

When you start up the mountain, you are going strong, but near the top you start to fade. You slow down from 20 km/h to 6 km/h in 6 minutes. What is your velocity (acceleration)?

(See Appendix for solution.)

12.6 Summary

- Linear motion is the motion of an object in a straight line.

- Linear motion is described by three quantities—speed, velocity (speed + direction), and acceleration.

- Speed is defined by the distance traveled divided by time: $s = d \div t$

- Velocity is defined by the distance traveled in a particular direction divided by time: $\mathbf{v} = d \div t$

- Acceleration is defined by the change in speed (or velocity) divided by the change in time: $a = \dfrac{\Delta \mathbf{v}}{\Delta t}$

- Mathematics is an essential tool for doing physics.

Chapter 13 Non-Linear (Curved) Motion

13.1	Introduction	133
13.2	Projectile Motion	133
13.3	Circular Motion	136
13.4	Summary	139

Physics

13.1 Introduction

In the last chapter we looked at how linear motion is defined by speed, velocity, and acceleration. We also looked at the equations for calculating speed and acceleration and how these equations work.

But what happens when motion is not linear? How can you calculate the speed of an object whose path curves as it moves? What happens when a ball rolls? Does the speed, velocity, or acceleration change? What happens when the wheels of a car rotate or when a cannon ball goes up and then down again?

In this chapter we will take a look at non-linear motion (curved motion), or the motion of objects that don't follow a straight line. One type of non-linear, or curved, motion includes the motion of any object (cannon ball, baseball, bullet, etc.) that follows a curved path. Another type of non-linear motion includes the motion of objects that rotate, or move in a circular pattern, like a car wheel, bike gear, or airplane propeller.

13.2 Projectile Motion

When you throw a basketball through a hoop, or a football down a field, or launch a pumpkin from a catapult, you are projecting a projectile. Projecting means "throwing, casting, or moving an object forward," and a projectile is the object that is being thrown, cast, or moved forward.

The motion of a projectile is described by its trajectory. A trajectory is the path the projectile takes at each instant in time once it is launched, thrown, or cast forward. When a ball or shot put is thrown in the air, the trajectory is non-linear (curved). In both of these situations, the projectile is launched into motion and is pulled toward Earth by gravity, resulting in a curved path.

Recall from Chapter 12 that linear motion can be described by speed, velocity, and acceleration. For an object traveling with linear motion, speed and velocity are often used interchangeably. As the object remains traveling in a straight line, the velocity need not change because the direction does not change. However, for non-linear (curved) motion the direction does change, and velocity and speed can no longer be interchanged.

The velocity for non-linear motion can be broken down (divided) into two components—a vertical component and a horizontal component. The vertical component describes the speed of the projectile directly perpendicular to the Earth's surface, and the horizontal component describes the speed of the projectile directly parallel to the Earth's surface. The total velocity is the sum of the vertical and the horizontal components of the velocity.

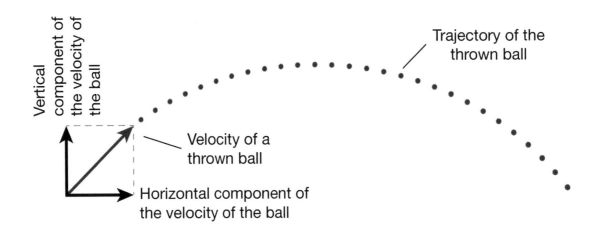

As the ball travels completing the trajectory, the velocity of the vertical component changes as gravity pulls the ball toward Earth. However, the horizontal component stays the same unless there is air resistance or some other force pushing the ball backwards. The total velocity changes as the ball travels through the arc.

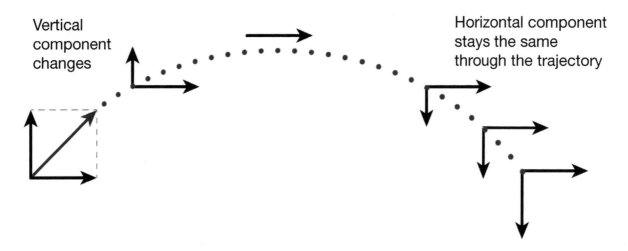

The distance a projectile will travel depends on the initial speed and the angle at which it is projected (direction). If the ball is thrown at too high an angle, it goes up and back down without going very far. If the ball is thrown at too shallow an angle, it reaches the ground before it has had a chance to travel very far. However, if the ball is thrown at a 45 degree angle, it has just the right combination of speed and direction to travel the farthest!

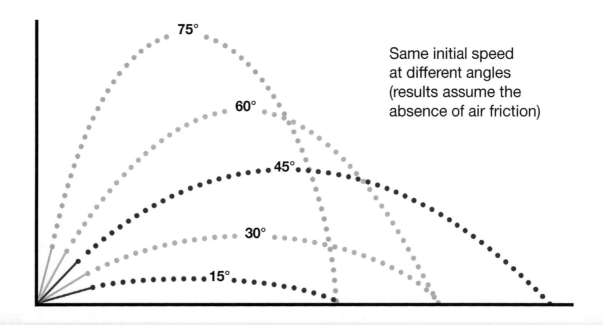

136 Exploring the Building Blocks of Science: Book 6

How can knowing about projectiles, angles, and speed help in the Punkin Chunkin competition? By understanding the physics, you can know that a pumpkin launched as a projectile will follow a curved trajectory. And knowing that a curved trajectory has both vertical and horizontal velocities, you can pick a path that predicts the distance the pumpkin will travel. You can also pick a launching mechanism that produces a high initial velocity that can be directed at a 45 degree angle. By applying your knowledge of physics, you can win the competition!

13.3 Circular Motion

Have you ever played on a merry-go-round? If you have, you probably noticed that when you push the merry-go-round, you run in a circle, and when you hop on and sit on the merry-go-round, it spins you in a circle. Moving in a circle is a special type of non-linear, or curved, motion called circular motion.

Circular motion has two different types of speed—tangential speed and rotational speed.

Physics—Chapter 13: Non-linear (Curved) Motion

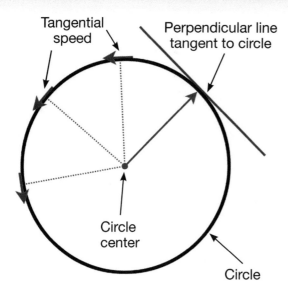

Tangential speed is the distance you travel along the path of the circle (tangential to the circle) over a given time. A tangent is a perpendicular line that intersects any point on a circle. Every point on the circle has a tangent. One way to think about tangential speed is to imagine an arrow traveling around the outer edge of the circle. It has tangential speed as it travels around the circle.

You may also have noticed that if you sit on the outside edge of the merry-go-round, you go faster than if you sit in the center. When you are farther from the center of a circle, the distance you travel is greater than when you are closer to the center of the circle so you have to go faster to make one complete revolution. Recall that speed is simply the distance traveled divided by time, so it makes sense that the tangential speed is greater farther from the center of the circle than it is closer to the center.

You Do It!

Physics Math Problem 3

Is there any tangential speed at the center point of the circle?

(See Appendix for solution.)

Rotational speed (also called angular speed) is determined by how many rotations a circle makes around its central axis in a certain amount of time. Rotational speed is the same both farther away and closer to the center of the circle. Rotational speed is often expressed as revolutions (rotations) per minute (RPM). Some cars have a gauge on the dashboard that measures the RPM of the engine. This gauge tells the driver how fast the engine is rotating the crankshaft of the motor.

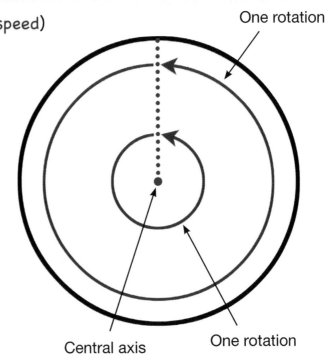

Some bicyclists use a meter called a cadence meter that measures how fast they are rotating the pedals. Cadence meters measure the RPMs of the pedals and help cyclists determine which gear to use to get the most speed for the least effort.

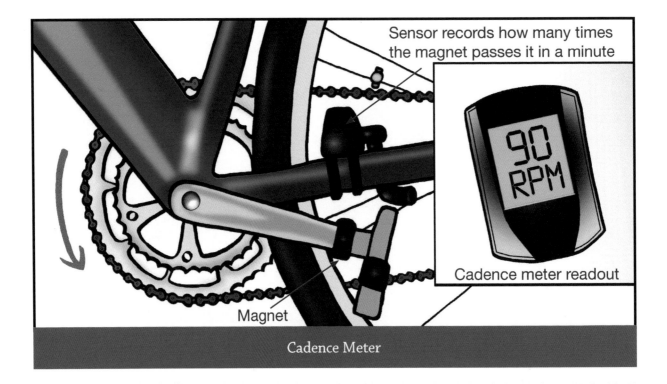

Cadence Meter

> ## You Do It!
>
> ### Physics Math Problem 4
>
> Place your finger on a rotating disk, such as a spinning bicycle wheel, other wheel, or a toy spinning top that has a flat upper surface, and let your finger move with the disk. Observe the difference between rotational speed and tangential speed. Place your finger on the outside of the rotating object and observe how far your finger travels in one revolution. Now place your finger on the inside of the rotating object and observe again how far your finger moves in one revolution. In both cases your finger traveled one rotation, so the rotational speed is the same.
>
> 1. Which position (closer or farther out) has more tangential speed?
> 2. Which position (closer or farther out) has more rotational speed?
> 3. What happens when you place your finger directly in the center?
>
> (See Appendix for solution.)

13.4 Summary

- Non-linear motion (curved motion) is any motion that does not travel in a straight line.

- A projectile is any object that is being thrown, cast, or launched.

- A trajectory is the path a projectile travels.

- The distance a projectile will travel depends on the angle and the initial speed at which it is projected.

- Circular motion is non-linear motion (curved motion) that travels in a circle.

- Circular motion is defined by both tangential speed and rotational speed.

Chapter 14 Technology in Geology

14.1	Introduction	141
14.2	Hand Tools	141
14.3	Electronic Tools	143
14.4	Other Tools	145
14.5	Satellites	146
14.6	Summary	147

Geology

Geology—Chapter 14: Technology in Geology

14.1 Introduction

Geologists use a variety of tools to study Earth. A hundred years ago geologists had only a few tools available. This limited what early geologists were able to study. But today, modern geologists use many different tools to measure, dig, monitor, and explore the Earth. Today's modern geologists use hand tools, electronic tools, and a variety of other tools.

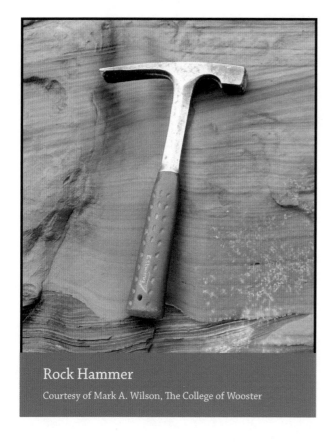

Rock Hammer
Courtesy of Mark A. Wilson, The College of Wooster

14.2 Hand Tools

Probably the most common hand tool used by geologists is the rock hammer or handpick. A rock hammer is a lightweight hammer used to break apart rocks or pry loose rock and soil away from other rocks. A rock hammer usually has a blunt end for hammering and a chiseled end for prying. Rock hammers vary in size and weight but usually are small enough and light enough to carry in a backpack. Sometimes when geologists photograph a rock sample, they will indicate the size of the sample by putting their hammer in the photo.

Another type of hand tool used by geologists is the crack hammer. A crack hammer is a small sledgehammer with two blunt ends, and it is heavier than a rock hammer. Crack hammers are used to break apart larger rocks.

The compass is another hand tool often used by geologists. In order to observe Earth's features, geologists must often travel outdoors in undeveloped wilderness areas. A compass can help a geologist navigate the terrain and find the road back to civilization after a long hike!

Along with a compass, geologists commonly use several different types of maps. Using different types of maps is a good way to understand a region.

City and road maps show cities, city landmarks, and roads and can be used to study geological features within and between populated areas.

Topographic maps give geologists an idea of how high or low an area might be. A topographic map has contour lines that show the shape and elevation of an area. When the lines are close together, it means the area is steep. When the lines are far apart, it means the area is flat or less steep.

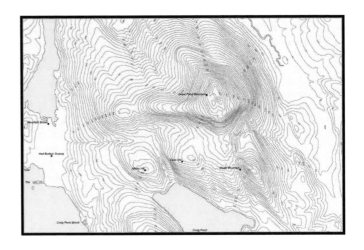

Climate maps give information about the climate of a particular region. A climate map shows the amount of precipitation, or how much rain or snow an area receives. Climate maps also show differences in temperature from region to

Geology—Chapter 14: Technology in Geology 143

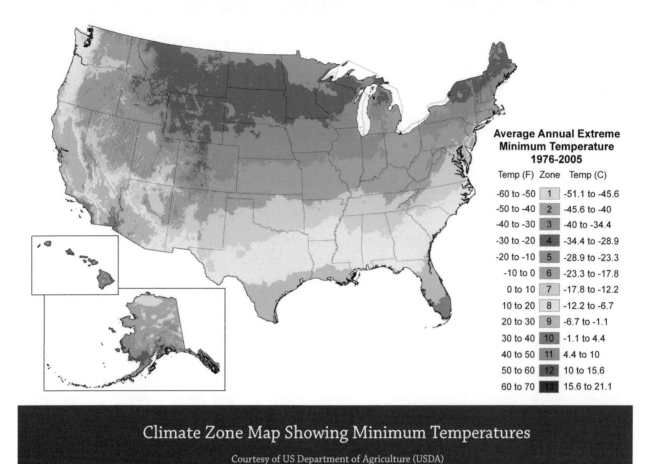

Climate Zone Map Showing Minimum Temperatures
Courtesy of US Department of Agriculture (USDA)

region. Climate maps are divided into climate zones accompanied by a key that shows which areas are dry or moist, hot or cold.

14.3 Electronic Tools

There are many modern electronic tools geologists can use to study the Earth. For example, imagine going into a dense forest, far from cities and roads, to study a particular type of rock or geological formation. If you are unfamiliar with the area, and if the maps are not very reliable, it might be easy to get lost. A modern tool you can

use to find your way is a GPS or global positioning system. A GPS is a device that uses satellite information to provide the location and time on or near the Earth. A GPS calculates position by precisely timing signals sent by satellites high above the Earth. The satellite sends messages to the GPS, and based on these signals, the GPS can calculate an exact location.

A GPS is handy for getting to places above ground, but what if you wanted to know what is below ground? One way to explore features below ground is to dig a hole, but holes can be dug only so far and digging can disrupt critical features a geologist might want to study. To "see" below the surface of the Earth, geologists can now use GPR, or ground penetrating radar. GPR is a method that uses high frequency radio waves to image below the surface. High frequency radio waves can be transmitted into the ground, and when the radio waves encounter an object, the waves will be reflected. A computer collects the reflected waves and creates an image of the objects below. GPR can be used to study ice formations, soils, groundwater, and even dinosaur bones!

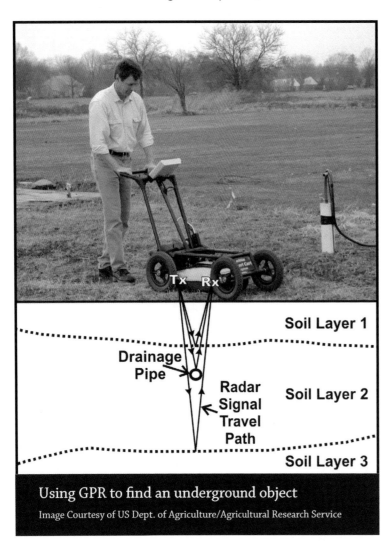

Using GPR to find an underground object
Image Courtesy of US Dept. of Agriculture/Agricultural Research Service

A GPR is useful for observing static objects—those that stay the same. But what about observing dynamic processes—those that change over time, like earthquakes and volcanic eruptions? To observe dynamic processes, geologists can use a seismometer

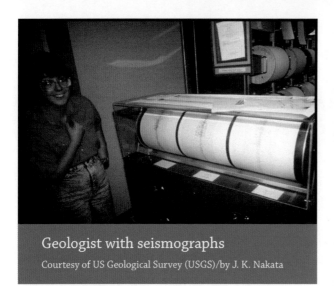

Geologist with seismographs
Courtesy of US Geological Survey (USGS)/by J. K. Nakata

The whole seismometer moves as the Earth it is attached to shakes, but the heavy mass does not move because of its inertia.

The recording device measures how far the rest of the seismometer has moved with respect to the mass.

Seismometer with seismograph
Courtesy of US Geological Survey (USGS)

or a seismograph. Both of these instruments measure motions in the ground. The words seismometer and seismograph come from the Greek word *seismos* which means "to shake or quake."

A seismometer is an instrument used "to measure shakes and quakes" and seismograph means "to draw shakes and quakes." Both instruments detect and measure the motion of the Earth's surface.

14.4 Other Tools

Sometimes a geologist might want to take a sample of rock from deep inside the Earth's surface. To do this a core drill can be used. A core drill is a drill specifically designed to go several hundreds to several thousands of feet into the Earth's surface. The core drill cuts a circular hole as it moves downward, creating a cylinder of rock, or core sample, that can be pulled out of the Earth. Core drills can be used for mineral exploration and to study the layers of rock deep in the Earth.

Drill bit used to take core samples of rock
Courtesy of US Geological Survey (USGS)

A rock and mineral test kit is another useful tool for geologists. Once a rock sample is collected, the rock and mineral test kit can be used to test for color, hardness, and how a sample reacts with certain acids. All of these features are helpful in identifying the types of minerals in the sample. A rock and mineral test kit is very useful for field geologists because it is small enough to be carried in a backpack.

14.5 Satellites

Satellites have become essential data gathering tools for geologists. In space, a satellite is an object that orbits another object. For example, the Moon is a satellite of Earth because it orbits Earth. An artificial satellite is a machine that has been launched into space and put into orbit around Earth. To gather information about Earth, the United States and many other countries have satellites in orbit that collect data about such features as the atmosphere, climate, geodynamics, gravity, weather, the oceans, ice, groundwater, the Sun and its influence on Earth, and the magnetosphere.

In 1972 NASA launched the first of what are now called the Landsat satellites. Using over 40 years of this satellite imagery, scientists are able to track how Earth's surface changes both slowly and quickly over time. Natural events such as hurricanes, blizzards, volcanic eruptions, earthquakes, and forest fires can be studied. The information collected can be used to identify the effects of natural events on Earth's surface and how these events affect people and the environment. The impact of human caused changes to the Earth, such as pollution, population growth, and deforestation, can also be studied, leading to new ideas for ways to protect the environment.

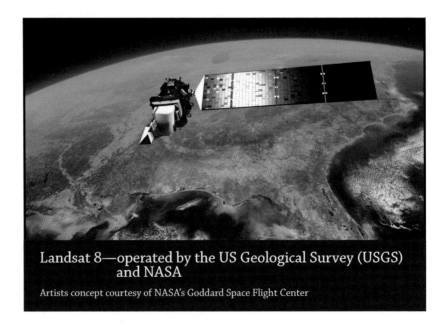

Landsat 8—operated by the US Geological Survey (USGS) and NASA

Artists concept courtesy of NASA's Goddard Space Flight Center

The SMAP (Soil Moisture Active Passive) satellite launched by NASA in 2015 measures and maps the amount of water in the top 5 cm (2 inches) of soil around the world. This will help scientists better understand the water cycle and how storms, drought, and the changes of season affect moisture in the soil, flooding, and the growing of crops. It will also provide data for studies of weather, climate, and the carbon cycle.

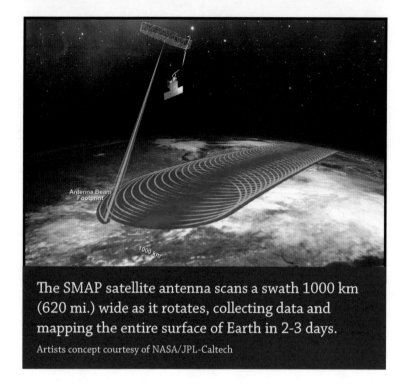

The SMAP satellite antenna scans a swath 1000 km (620 mi.) wide as it rotates, collecting data and mapping the entire surface of Earth in 2-3 days.
Artists concept courtesy of NASA/JPL-Caltech

14.6 Summary

- Geologists use many different tools to measure, dig, monitor, and explore the Earth.

- Rock hammers and crack hammers are hand tools geologists use to collect rock samples.

- Geologists use various types of maps, such as topographic maps and climate maps, to help them understand the region of the Earth being studied.

- Electronic tools are helpful in finding locations, measuring vibrations in the ground, and visualizing objects below the surface of the Earth.

- Satellites are important tools for gathering data about Earth.

Chapter 15   Earth's Spheres

15.1	Introduction	149
15.2	The Spheres of Earth	150
15.3	Connecting the Spheres	153
15.4	A Delicate Balance	156
15.5	Summary	158

Geology

Geology—Chapter 15: Earth's Spheres 149

15.1 Introduction

As you go about your day, you may not think much about all the different parts that make up the Earth. If you live in a house or an apartment, your days are likely illuminated with artificial light rather than just the Sun. If you live in a city, you probably can't see many stars from your bedroom window at night. If you travel from home by car, you can't feel the ground beneath your feet or smell the rain outside. If you shop at  the grocery store, you may buy apples, oranges, and bananas, but you probably don't think about the trees, dirt, bugs, and worms that are needed for making the apples, oranges, and bananas grow. And unless you enjoy using a compass to find your way while hiking, you probably don't think about the magnetic field that surrounds the Earth.

However, if you live on a farm or in the woods or wilderness, you are likely to be much more familiar with the different parts that make up Earth. If you've lived long enough on a farm or in the woods or wilderness, you can probably predict rain by the shape of the clouds and know when the soil is just right for planting and which plants do well when planted in the heat of summer and which do better planted when it's colder. You notice that the length of the day varies with the seasons, and you likely know when to expect the first frost so you can harvest crops before they freeze. You may know when the nearby lake will be full of young fish, tadpoles, and ducklings and when it will be frozen over with a layer of ice thick enough to skate on. You probably also know how the rain affects the growth of plants, how animals live by eating plants, and how both plants and animals change the Earth's soil.

In today's modern world, many of us have lost touch with how water, air, ground, plants, animals, and the magnetic field shape and govern Earth's habitats. These features (air, water, ground, plants and animals, and magnetic field) are all part of an interconnected system of spheres that make life on Earth possible.

15.2 The Spheres of Earth

Earth is a system. A system is a set of different parts that work together as a whole. The different parts that make up Earth surround the planet, and because Earth is almost spherical in shape, these parts that surround Earth are referred to as spheres. Although Earth's spheres are sometimes categorized differently, in this textbook we will describe the Earth as being made of five different spheres: the geosphere, biosphere, hydrosphere, atmosphere, and magnetosphere.

The geosphere makes up the rock part of Earth—the crust, mantle, and core. The crust is the solid outer layer of Earth that is made of rocks and minerals and is the part of the geosphere where plants and animals make their home. The geosphere is constantly being changed by earthquakes and the eruption of volcanoes and by the erosion of rocks by wind and water. Deep in the interior of the geosphere, the swirling of molten rock in the outer core creates Earth's magnetic field.

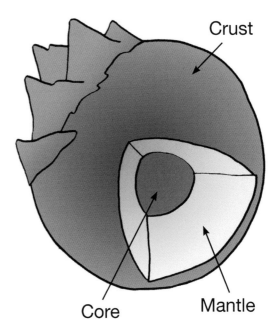

Geology—Chapter 15: Earth's Spheres **151**

The atmosphere is made up of the air and tiny particles that surround the Earth. The atmosphere is an ever changing mixture of gases and particles that create a variety of weather events including hurricanes, clouds, cyclones, dust and sand storms, wind, typhoons, and blizzards. The atmosphere is also responsible for long-term fluctuations in climate.

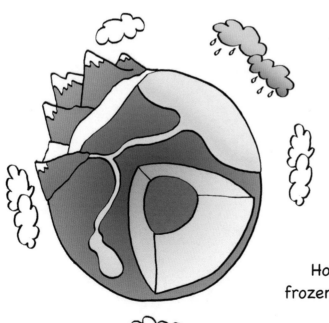

The hydrosphere makes up all the liquid and solid water on Earth. The hydrosphere includes the oceans, lakes, rivers, streams, ponds, groundwater, glaciers, water vapor in clouds, rain, snow, and icebergs. Sometimes water that is frozen and solid is considered separately from the hydrosphere and is called the cryosphere. However, in this text we will include frozen water as part of the hydrosphere.

The biosphere includes all the living things on Earth—all the life on land, in the oceans, in the dirt, underneath glaciers, and near high temperature thermal vents. All living things from the biggest land and sea creatures to the tiniest microscopic organisms are part of the biosphere.

The magnetosphere surrounds the Earth in space. The magnetosphere contains Earth's magnetic field and is formed from the interaction of the magnetic field with solar winds. The magnetosphere protects the Earth from harmful ions and solar winds coming from the Sun, and it can change shape as a result of solar variations.

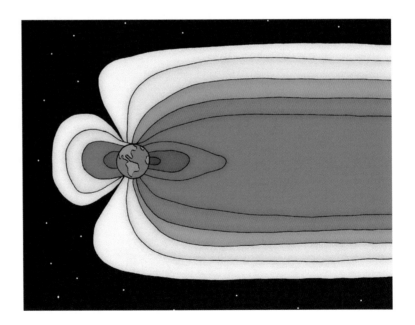

15.3 Connecting the Spheres

All of Earth's spheres are connected to each other. The atmosphere is connected to the hydrosphere, the geosphere, the biosphere, and the magnetosphere. The hydrosphere is connected to the atmosphere, the geosphere, the biosphere, and the magnetosphere. All of the spheres connect to each other. A scientist who studies climate must consider influences from the atmosphere, hydrosphere, biosphere, geosphere, and magnetosphere just as a scientist

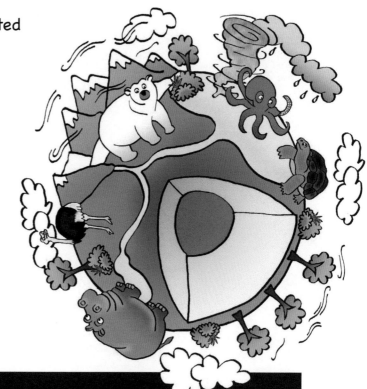

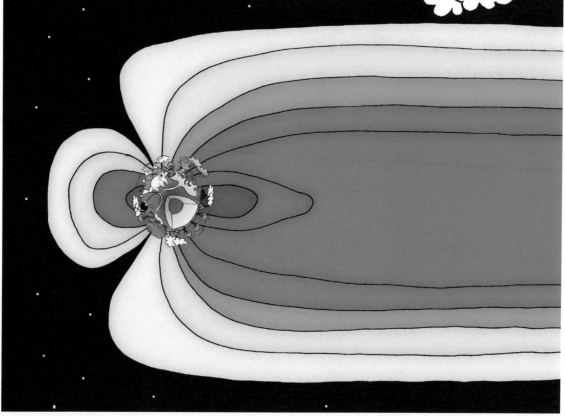

who is studying plants and animals must consider the geosphere, atmosphere, hydrosphere, and magnetosphere in addition to the biosphere.

It is not always easy to determine how the spheres interact with one another. Although some effects can be immediately seen, others may take years or decades to notice. For example, a typhoon caused by the interaction of the atmosphere and hydrosphere may wipe out the entire habitat (biosphere) of one part of an island, and the change in plant and animal populations will be immediately evident. However, if that typhoon created a channel through soft soil (geosphere) that continues to erode with every rain storm, a river or gorge may develop over years or decades.

Recall from Section 15.2 that Earth's spheres can be thought of as a system made up of different parts. The nature of Earth as a system is obvious when we notice how the different parts (spheres) interact with one another. Many of the interactions between the spheres occur in complicated cycles.

Some major cycles that connect the spheres include the energy cycle, the water cycle, the rock cycle, the carbon cycle, and the nitrogen cycle. Understanding how each of these cycles interacts with the different spheres of Earth is like putting together a huge, complicated puzzle. It takes many scientists working simultaneously on different areas of research to understand what the puzzle pieces are, how they work, and how they interact with each other.

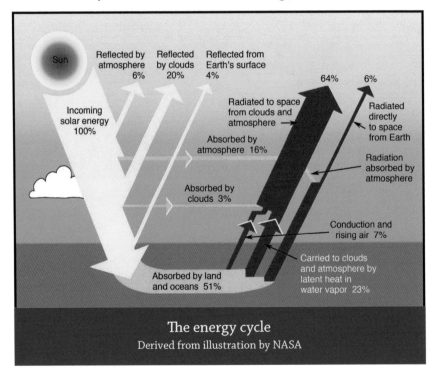

The energy cycle
Derived from illustration by NASA

Geology—Chapter 15: Earth's Spheres

To better understand how Earth's spheres interact, scientists have been taking photographs of Earth's surface from satellites. Using satellite images scientists can observe how changes in weather patterns alter coastlines, forested areas, glaciers, and waterways. Satellite imagery can also help scientists understand how changes in glaciers, forested areas, waterways, and coastlines can change the climate in different parts of the world. Also, changes to the spheres from natural disasters and human activity can be observed and analyzed.

The satellite images below show how flooding can affect the geosphere, biosphere, and hydrosphere and how fire can affect the atmosphere and biosphere.

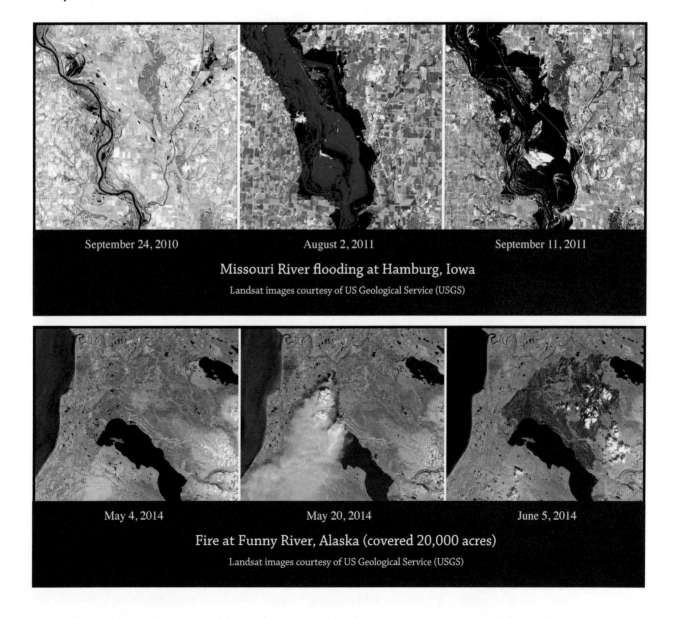

September 24, 2010 — August 2, 2011 — September 11, 2011

Missouri River flooding at Hamburg, Iowa

Landsat images courtesy of US Geological Service (USGS)

May 4, 2014 — May 20, 2014 — June 5, 2014

Fire at Funny River, Alaska (covered 20,000 acres)

Landsat images courtesy of US Geological Service (USGS)

15.4 A Delicate Balance

Because Earth's spheres are connected to each other through complex cycles, there is a delicate balance between them. Earth has to maintain this balance in order for living things to grow and be healthy. Activities that disrupt one or more of these interconnected cycles can throw Earth's ecosystems out of balance.

For example, it is well know that burning fossil fuels causes air pollution in the atmosphere, lowering the pH of rain, making the rain acidic. Acid rain in the hydrosphere can cause damage to the leaves of trees and other plants in the biosphere and can make the soil of the geosphere more acidic. Acid rain can also speed up the rate of decomposition in the soil. Decomposition of leaves and other organic matter releases carbon dioxide into the soil, and the CO_2 in the soil is gradually released into the atmosphere. Rapid decomposition can result in greater quantities of carbon dioxide being produced and released from the soil into the atmosphere. This adds to the gases in the air that prevent heat from escaping from Earth, causing a warming effect which leads to further changes in the atmosphere, hydrosphere, biosphere, and geosphere.

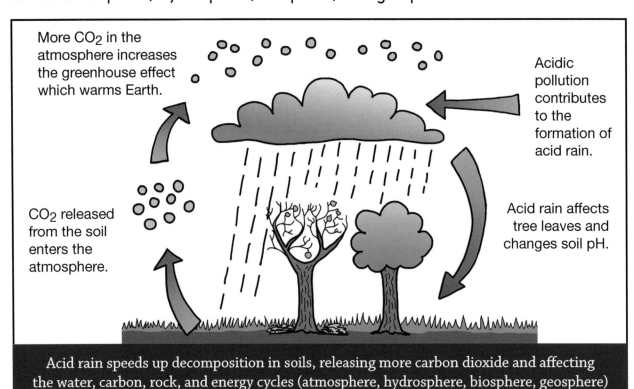

Acid rain speeds up decomposition in soils, releasing more carbon dioxide and affecting the water, carbon, rock, and energy cycles (atmosphere, hydrosphere, biosphere, geosphere)

Also, natural disasters can effect the balance of Earth's spheres. A massive volcanic eruption can put so much ash in the atmosphere that the ash particles can block some of the Sun's light from the Earth, causing cooling. Increased rainfall can also result as water droplets form on ash particles. Ash and lava that fall on the geosphere can kill plants and animals and change soils and landforms.

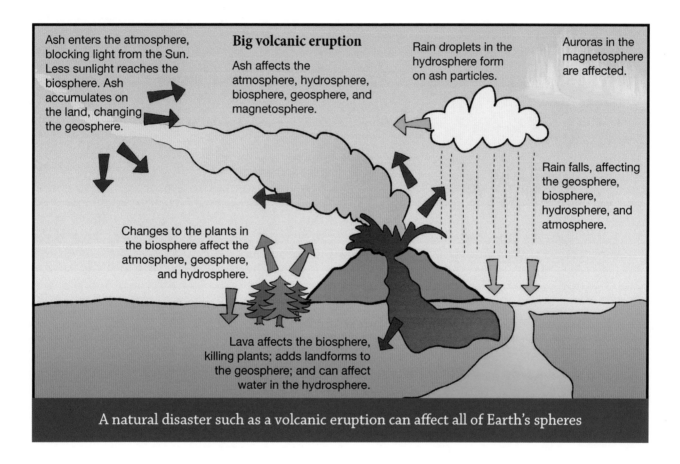

A natural disaster such as a volcanic eruption can affect all of Earth's spheres

There are many properties of Earth that must remain balanced in order for humans to thrive. Humans require a narrow temperature range, a certain amount of oxygen to breathe, a certain amount of water to drink, adequate amounts of fertile land to grow crops and raise animals for food, and materials for making clothing and building shelter.

Over the last several decades scientists have become increasingly concerned about how human activities may be disrupting the delicate balance of Earth's interconnected systems. For example, scientists have been observing changes in the amount of carbon dioxide in the atmosphere and changes in Earth's temperatures. Carbon dioxide is called a greenhouse gas when it enters the atmosphere because it affects how Earth regulates its temperature in a way that is similar to how a greenhouse regulates temperature. Earth's average temperature is the result of an intricate balance between heating by the Sun and the escape of heat from the atmosphere into space. Greenhouse gases play a role in this balance. We'll learn more about the atmosphere and greenhouse gases in Chapter 17.

Studying Earth's spheres and how they fit together and interact can help scientists to come up with new ways for predicting changes to the Earth from natural and human causes and develop new ideas for helping people protect and restore environments.

15.5 Summary

- Earth can be described as a system of interconnected spheres that include weather, plants and animals, rocks and dirt, water, energy, and Earth's magnetic field.

- Earth's spheres are commonly referred to as: the atmosphere, the hydrosphere, the geosphere, the biosphere, and the magnetosphere.

- Earth's spheres are interconnected, and changes in one sphere can result in changes in another sphere.

- Earth must maintain a delicate balance between the spheres to continue to support life.

Chapter 16 The Geosphere

16.1	Introduction	160
16.2	Minerals and Elements	162
16.3	Using Volcanoes To See Inside Earth	163
16.4	Using Earthquakes To See Inside Earth	166
16.5	How Hot Is the Core?	170
16.6	Summary	171

Geology

16.1 Introduction

If you take a walk through the forest, you might notice birds sitting in the trees, small plants growing from the soil, mushrooms bursting out of rotten logs, and squirrels gathering food for the winter. All of the living things you see are supported by the rocks, soil, and minerals beneath the forest floor. Without a solid surface, plants could not grow and animals could not collect the food they need for survival. The rocky part of Earth is called the geosphere and is made of rocks, minerals, soils found on the surface, and molten rock, or magma, found deep below the surface.

Humans live on the uppermost part of the geosphere. Although people can travel in the sky to get to another city or traverse the oceans in a boat, all of human life is supported by the rocky geosphere. Humans build houses from materials found on Earth's geosphere, grow plants and raise animals on Earth's geosphere, and use rocks and minerals from the geosphere to create cars, airplanes, and trains. Humans also extract gas and coal from the geosphere to fuel factories and power boats and even use nuclear energy obtained from rocks to provide electricity to entire cities. It is an important task for geologists to understand how the geosphere is made, what parts it has, and how to best use the resources it contains.

Recall that the geosphere is layered with a solid outer crust, a layer below the crust called the mantle, and the core at the very center of Earth. The mantle can be further subdivided into sublayers called the lithosphere, asthenosphere, and mesosphere. The core has two sublayers—the outer core and the inner core. These divisions indicate differences in composition, with some layers being hard and solid and other layers being soft or liquid.

The crust holds all the rocks, soils, minerals and bodies of water where organisms live. The crust is often grouped with the lithosphere; however, in this text we will denote the lithosphere and crust as separate layers, with the lithosphere being the upper layer of the mantle. The lithosphere lies just below the crust, and although no one has drilled deep enough to take rock samples, scientists find evidence to support the idea that the lithosphere is a hard, rocky layer divided into huge pieces called tectonic plates.

The asthenosphere sits just below the lithosphere. Using seismic data, scientists have concluded that the asthenosphere is made of a soft, putty-like material made of magma. (We will learn more about how seismic data can be used to see inside the Earth in Section 16.4.) Below the asthenosphere is the mesosphere. Scientists have found seismic evidence that suggests that the mesosphere is more solid than the asthenosphere. Below the mesosphere lie the outer core and the inner core. Seismic data support the idea that the outer core is liquid and the inner core is solid.

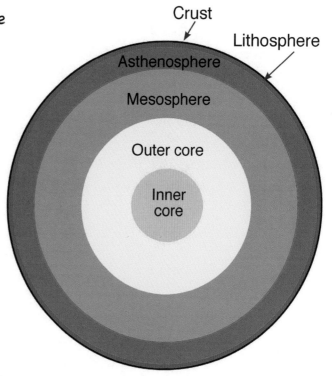

The geosphere interacts with Earth's other spheres, and Earth's other spheres interact with the geosphere. The crust is modified by the biosphere as plants and animals move dirt and soil to grow or create homes; the hydrosphere as rains dig into Earth's crust creating rivers, ponds, and oceans; and the atmosphere as winds, hurricanes, and tornados alter the outline of beaches, break down rocks, and move dirt and soil to new locations. The geosphere creates the magnetic field that is part of the magnetosphere. Although in this chapter we will look at the geosphere separately from the other spheres, it is important to remember that all of Earth's spheres are connected and interact with each other.

16.2 Minerals and Elements

If we walk through the woods, along a desert road, or cross a stream barefoot, we will walk on rocks. Rocks are everywhere—in the forest, at the top of mountains, and at the bottom of the ocean. Where do rocks come from?

We know that although rocks are made in different ways, all rocks begin as minerals. Recall from Book 5, Chapter 15 that minerals contain the elements oxygen, silicon, aluminum, iron, calcium, sodium, potassium, magnesium, and others; form crystals with different shapes; and have different colors and different degrees of hardness. Although there are more than 4000 known minerals, only a handful of minerals make up Earth's crust and inner layers. These minerals are called rock-forming minerals.

Rock-forming minerals make up approximately 99.7 percent of the mass of the crust. Most of the rocks that make up the Earth's crust are silicates composed mainly of the elements oxygen and silicon. However, iron makes up about 35 percent of the entire Earth because a large part of Earth's core is iron.

Scientists are able to study the rocks and minerals of Earth's crust directly by making observations of rocks and rock formations on the surface. They can also study core samples obtained by drilling and rock samples taken from mines. By analyzing these rock samples, scientists can learn more about how rocks are formed deep below the surface. However, the deepest drill hole is only 12 km (7.5 mi.) long and not nearly deep enough to sample all the way through the Earth's crust to the mantle below. Studying volcanic activity and earthquakes allows geologists to investigate the materials far below Earth's surface.

Geology—Chapter 16: The Geosphere

16.3 Using Volcanoes To See Inside Earth

How do scientists know what Earth's interior is made of if we can't sample rocks and minerals below the upper part of the crust? One way scientists study rocks and minerals deep in the Earth is to observe what happens when volcanoes erupt and to study the lava, rocks, and ash that come from the volcanoes.

1. A volcanic eruption in Hawaii Courtesy of Hawaiian Volcano Observatory (HVO)/USGS by J. D. Griggs
2. Setting up a GPS system to measure deformation of ground (getting higher or lower), Mt. St. Helens, Washington State Courtesy of USGS by Mike Poland; 3. Collecting volcanic ash samples in Alaska Courtesy of Alaska Volcano Observatory (AVO)/USGS by Kristi Wallace; 4. Taking a sample of lava, Kilauea, HI Courtesy of HVO/USGS; 5. Taking gas samples at the Cookie Monster skylight, HI Courtesy of HVO/USGS by J. D. Griggs

Gneiss (metamorphic) Granite (igneous)
Gneiss photo courtesy of Huhulenik - CC BY 3.0 via Wikimedia

In a volcanic eruption, magma from within the Earth is brought to the surface. In addition, pieces of rocks formed inside the Earth are often ejected when they are torn loose and carried by the magma as it pushes towards the surface with tremendous force. Granite, an igneous rock, and gneiss, a metamorphic rock, are the most common types of rock ejected by volcanoes.

Rare gems such as diamonds have been found in rocks that have been ejected during volcanic eruptions. Diamonds are made of pure carbon and most often form more than 150 km (93 mi.) below Earth's surface. Volcanoes also bring up chunks of peridotite, an igneous rock made mostly of the minerals olivine and pyroxene. Peridotite may also contain garnets, some of which can be used as gemstones. Laboratory experiments show that garnet peridotite forms at depths greater than 50 km (31 mi.).

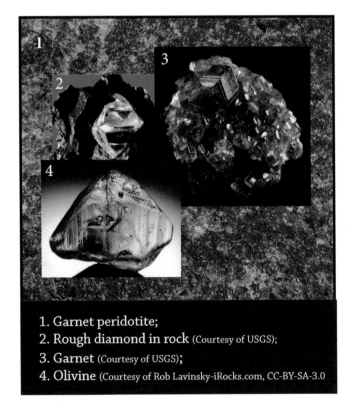

1. Garnet peridotite;
2. Rough diamond in rock (Courtesy of USGS);
3. Garnet (Courtesy of USGS);
4. Olivine (Courtesy of Rob Lavinsky-iRocks.com, CC-BY-SA-3.0

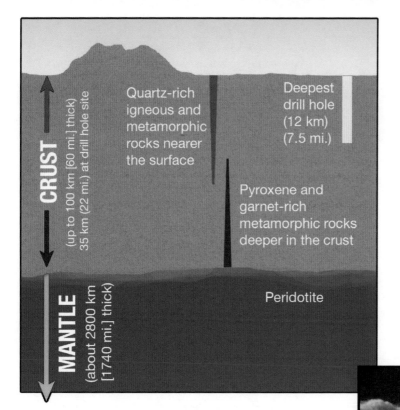

Information gathered from volcanoes has given scientists an idea about what forms below Earth's surface, with quartz-rich rocks forming nearer the surface of the crust and pyroxene and garnet peridotite in the lower part of the crust closer to the mantle.

Geologists also study the makeup of volcanic ash to learn more about the interior structure of Earth. Volcanic ash is formed when a volcano erupts explosively. Magma contains dissolved gases that can cause the magma to explode if the gases expand and escape violently into the atmosphere. Solid rock can also be exploded when the force of the escaping gases shatters the rocks.

Ash from Mt. St. Helens, WA collected 39 km away in Idaho
Courtesy of Cascades Volcano Observatory/USGS

Eruption plume of gases and ash Mt. St. Helens, Washington State
Courtesy of Cascades Volcano Observatory/USGS

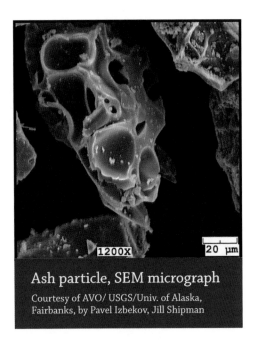

Ash particle, SEM micrograph
Courtesy of AVO/USGS/Univ. of Alaska, Fairbanks, by Pavel Izbekov, Jill Shipman

Volcanic ash isn't soft like the ash from a wood fire. Instead, it is made of hard, sharp bits of rock that are the size of grains of sand or smaller. Because the force of the explosion thrusts the ash up into the atmosphere, winds can carry the tiny particles far from the volcano—as much as thousands of kilometers away! By studying the composition of the ash particles, geologists can learn more about what the interior of Earth is made of.

16.4 Using Earthquakes To See Inside Earth

If you live in an area of the globe that's prone to earthquakes, you might have awakened in the middle of the night to a deep rumbling sound and felt your house being shaken by an earthquake. Earthquakes happen in different parts of the world and can be devastating for people living near the origin, or epicenter, of the earthquake. They can cause extensive damage to homes, schools, city buildings, farmland, and other structures that exist on Earth's surface.

A building collapsed on a car during an earthquake, Loma Prieta, CA
Courtesy of USGS by J. K. Nakata

However, earthquakes are very useful for finding out more about Earth's layers. Earthquakes create waves that are caused by the movement of the crust and lithosphere, and these waves can be used to create images of Earth's interior.

Recall that waves are the transfer of energy through a medium such as air, water, wood, rocks, magma, and other materials (see Book 4, Section 10.4). When you throw a pebble into a pond or strike a tuning fork against a surface, you are creating waves. Also recall that molecules in the medium through which a wave travels are displaced, but the molecules are not carried along with the wave. It is only the energy that moves as the wave moves.

When the tectonic plates of Earth's crust move, slide, or bump up against each other, stresses build up as the rough edges of the plates stick together, but the plates themselves continue to move. An earthquake occurs when the stresses become too great and the plates suddenly lurch as the edges release their hold on each other. An earthquake starts at the point where the release between two plates occurs (the epicenter), and the energy this sudden movement creates travels through the surrounding rock in the form of waves.

To measure the waves generated by an earthquake, geologists use a seismometer. Recall from Chapter 14 that a seismometer is a very sensitive instrument that can detect and amplify waves on Earth's surface, in the body of the crust, and below the crust.

How can earthquake waves tell us about Earth's interior? It turns out that the way an earthquake wave travels through the Earth's crust and also through Earth's lower layers allows geologists to construct an image of Earth's interior. Earthquake waves can be used like an x-ray, giving scientists a picture of Earth's internal structure. By using seismometers stationed in many locations on Earth, scientists can determine how fast a seismic wave from an earthquake travels through the Earth from the epicenter of the earthquake to where it comes back to the surface in another location. Scientists can also determine the direction the wave is traveling. Combining the direction of the wave with its speed reveals the velocity of the wave.

Recall from Book 4, Chapter 10 that the velocity of a wave is dependent on the type of material the wave travels through. Knowing the velocity and type of seismic waves allows scientists to image the Earth's interior.

There are two types of waves that travel through Earth below the surface. These are classified as primary (P) waves and secondary (S) waves. Both P and

S waves are called body waves because they move through material that is below the surface, or in the body of, the Earth. By measuring these waves, we get information about Earth's inner layers. Because these waves are confined by the pressure of the overlying rock, the rock moves only a small distance as the waves travel through, and therefore P and S waves are not as damaging as surface waves.

A P wave is a longitudinal wave where material is displaced parallel to the direction the wave is moving in a squeezing and stretching fashion.

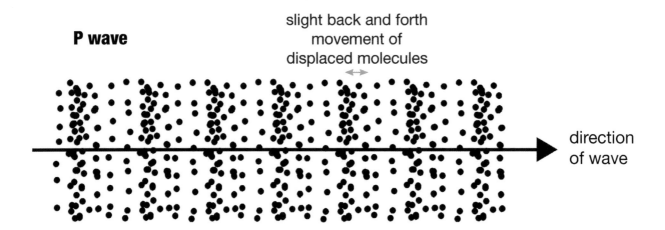

An S wave is a type of transverse wave where the material moves perpendicular to the direction the wave is traveling. P waves and S waves travel differently from each other through liquids and solids.

Molecules are displaced as a wave passes through them

P waves and S waves also reflect and refract as they travel, which tells scientists something about the boundaries between different types of materials. Recall from Book 4, Chapter 11 that when a wave encounters the boundary between two different types of materials, some of the energy may be reflected, creating a new wave, and some of the energy may be refracted, bending and changing velocity as it encounters a different type of material. By looking at how seismic waves reflect, refract, and change speed, scientists can determine where different types of rock may be located.

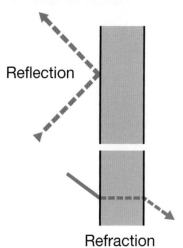
Reflection

Refraction

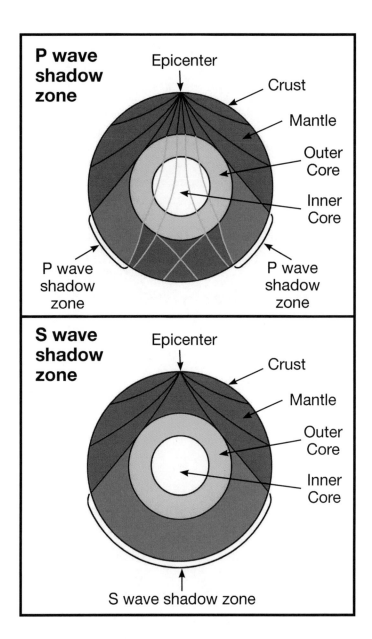

This illustration shows how P waves and S waves travel through Earth from the epicenter of an earthquake. The initial waves are caused by movements of tectonic plates that create an earthquake. The waves from the earthquake then pulse through Earth's layers. As the waves move, they are reflected and refracted when they encounter different materials. In some cases the waves are not able to travel through a material. This leaves areas where waves cannot be detected at the surface by seismometers, and these areas are called shadow zones.

Scientists theorize that an S wave shadow zone is created when the S waves from an earthquake hit the outer core and cannot travel through it.

Since S waves are unable to travel through liquid materials, scientists conclude that the outer core is most likely liquid.

P waves coming in contact with the liquid outer core behave differently from S waves. The P waves are refracted (bent) and slowed and create P wave shadow zones. The stopping of the S waves and the refraction of the P waves results in the different shadow zone areas.

We can see that the behavior of seismic waves helps scientists explore and identify the interior parts of Earth's crust, mantle, and core. There is much yet to be discovered about Earth's interior.

16.5 How Hot Is the Core?

How hot is the center of the Earth? If you've ever seen a volcano erupt and watched as hot lava consumed trees, buildings, and cars, you can imagine that the molten rock in the mantle of the Earth is very hot. But exactly how hot is the molten rock at the center of Earth?

The information gathered from earthquake seismic waves traveling through Earth gives scientists some very basic information about the temperatures of the inner and outer cores. Because the analysis of seismic waves can determine whether rock is solid or liquid, scientists can tell whether the rock is above or below its melting temperature. As we have seen, the movement of seismic waves suggests that the outer core is liquid and the inner core is solid.

Since the inner core is thought to be mostly iron that is under extreme pressure from the weight of all the materials above it, lab tests have been done that put iron under extreme pressure to get an indication of what core temperatures might be. The results of these tests combined with computer modeling have helped scientists arrive at estimated temperatures for the inner core.

Currently, the combined data gathered by various research methods suggests that the outer core is about 3000° to 5000°C (5400° to 9000°F). The inner core is thought to reach temperatures of about 6000°C (10,800°F), which is as hot as the surface of the Sun!

16.6 Summary

- The geosphere is the rocky part of Earth and is made of rocks, minerals, soils found on the surface, and magma found deep below the surface. The geosphere extends all the way from the surface of Earth to its very center.

- The Earth has layers that are divided into different sublayers.

- Lava, ash, and rocks ejected by volcanoes can be used to investigate what exists below Earth's crust.

- Earthquakes create waves that can be detected by seismometers and used to investigate Earth's layers.

- The temperature of Earth's inner core is believed to be about 6000°C.

Chapter 17   The Atmosphere

17.1 Introduction   173

17.2 Chemical Composition   173

17.3 Structure of the Atmosphere   175

17.4 Atmospheric Pressure   178

17.5 Gravity and the Atomosphere   180

17.6 The Greenhouse Effect   180

17.7 Summary   183

Geology

17.1 Introduction

We know that we need to have air to breathe in order to live. But what is air made of? How far does it extend above the surface of the Earth? Is all of the air the same? Why doesn't it float away into space?

The thin layer of gases that surrounds the Earth is called the atmosphere. The word atmosphere comes from the Greek word *atmos*,

which means "vapor" and refers to the gaseous state of matter. Without the atmosphere, life on Earth would not be possible. Earth's atmosphere contains the oxygen that animals need to breathe in order to live. It also protects the Earth and its inhabitants from getting too much energy from the Sun and helps keep the Earth's temperature from getting too hot or too cold. It carries the water that falls as rain and snow.

17.2 Chemical Composition

Earth's atmosphere is made up of several gases. Although oxygen is essential for life, it is not the most plentiful gas in the atmosphere. Instead, nitrogen makes up about 78% of the atmosphere, with oxygen at about 21%, and other gases including carbon dioxide, ozone, methane, and argon making up the remaining 1%. These percentages are for the "dry" atmosphere. But water vapor is also present in the atmosphere at an average of about 1% of volume.

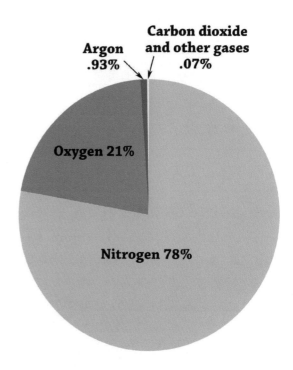

There are several features in Earth's atmosphere that allow life to exist and also protect it. These features include the amount of carbon dioxide, the ozone layer, water vapor, and the presence of oxygen.

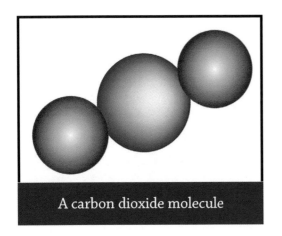

A carbon dioxide molecule

Carbon dioxide (CO_2) is a molecule made of one carbon atom and two oxygen atoms. It makes up less than 1% of the atmosphere. Carbon dioxide helps stabilize Earth's climate by trapping some infrared energy (heat) from the Sun which keeps Earth from freezing. However, too much carbon dioxide can trap too much infrared energy which can then cause the Earth to get warmer (see Section 17.6). Carbon dioxide is also essential to the process of photosynthesis in plants. Plants convert carbon dioxide to sugars to use for food. They then release oxygen into the atmosphere. Photosynthesis is the major source of oxygen on Earth.

Natural sources of atmospheric carbon dioxide include animal and plant respiration, by which oxygen and nutrients are converted into carbon dioxide and energy; ocean-atmosphere exchange, in which the oceans absorb and release carbon dioxide at the sea surface; and volcanic eruptions, which release carbon from rocks deep in the Earth's crust.

Man-made sources of carbon dioxide include the burning of fossil fuels for heating buildings, power generation, and transportation, as well as some industrial processes.

Ozone (O_3) is a molecule made of three oxygen atoms. Ozone is also an important gas in our atmosphere although it makes up less than .0001% of the total atmospheric gases. The ozone layer blocks most of the ultraviolet radiation from reaching the Earth's surface (see Section 17.3) but allows some of the incoming energy from the Sun to be trapped in the atmosphere. This process also slightly heats the

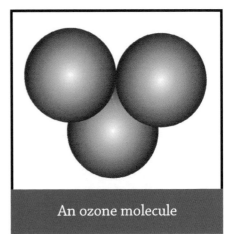

An ozone molecule

Earth, keeping temperatures mild and stable. Without the ozone layer, Earth would be uninhabitable.

Water vapor is also an important atmospheric gas. The atmosphere close to the Earth's surface contains up to 4% water vapor. This water vapor greatly affects our weather. Water vapor is actually the most important molecule for keeping our Earth warm. Water vapor is also responsible for the formation of clouds and rain that give plants and animals the essential water they need for life.

Finally, the most important atmospheric gas for animal life is oxygen. Without oxygen, life as we know it would not be possible. Animals breathe in oxygen and this oxygen is moved through cells and used for a variety of biochemical processes that make animal life possible.

17.3 Structure of the Atmosphere

Most of the mass of Earth's atmosphere exists within 20 kilometers (12 miles) of the Earth's surface. However, above these first 20 kilometers of air there are several atmospheric layers that function to protect the Earth's surface, absorb harmful radiation, and keep the Earth warm.

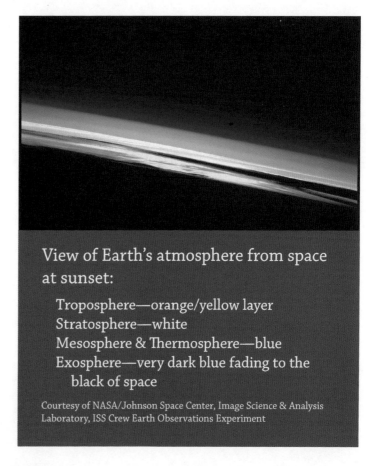

View of Earth's atmosphere from space at sunset:

Troposphere—orange/yellow layer
Stratosphere—white
Mesosphere & Thermosphere—blue
Exosphere—very dark blue fading to the black of space

Courtesy of NASA/Johnson Space Center, Image Science & Analysis Laboratory, ISS Crew Earth Observations Experiment

Scientists have divided the atmosphere into five layers—troposphere, stratosphere, mesosphere, thermosphere, and exosphere. These different layers are the result of differences in temperature, chemical composition, movement, and density of the gases in the atmosphere.

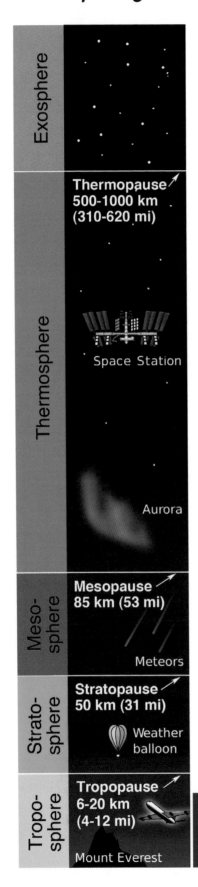

The top boundary of each atmospheric layer is called a "pause," so there is a tropopause, a stratopause, a mesopause, and a thermopause. A "pause" is an area where a change in temperature occurs, with the temperature either changing abruptly or reaching a minimum or maximum temperature.

The first layer above the Earth's surface is called the troposphere and is the layer we live in. The troposphere contains 78-80% of the Earth's atmospheric gases and extends to a height of about 6-20 kilometers (4-12 miles) from the Earth's surface. All of the weather we experience happens in the troposphere.

The layer above the troposphere is called the stratosphere. The stratosphere extends approximately 50 kilometers (31 miles) above Earth's surface. The air in the stratosphere is very dry with few gases and very little water vapor. Because there are fewer gases, this layer is more stable than the troposphere. Few, if any, clouds form in the stratosphere and airplanes fly in the lower stratosphere to avoid weather conditions that cause turbulence and bumpy flights.

The ozone layer is found at the lower part of the stratosphere. Ozone is crucial to life on Earth because ozone absorbs harmful ultraviolet radiation from the Sun. Ultraviolet radiation has a shorter wavelength than visible light and can damage living cells. Sunscreen and sunglasses help protect

Structure of the Atmosphere

Derived from illustration by National Weather Service, National Oceanic and Atmospheric Administration (NOAA)/Department of Commerce

people from the damage ultraviolet radiation causes. Without the ozone layer, organisms would suffer severe burns from ultraviolet radiation. The ozone layer also creates increased temperatures in the stratosphere.

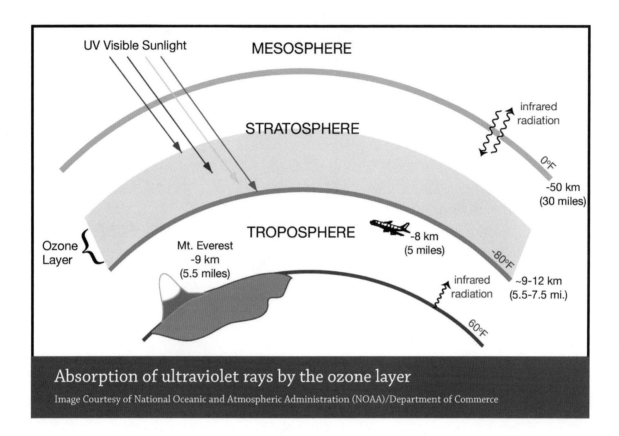

Absorption of ultraviolet rays by the ozone layer

Image Courtesy of National Oceanic and Atmospheric Administration (NOAA)/Department of Commerce

The mesosphere begins at an altitude of approximately 50 kilometers (31 miles) and reaches heights of about 85 kilometers (53 miles). In this layer the temperatures again begin to get cooler the higher you go. The top of the mesosphere is the coldest part of the atmosphere with temperatures of approximately -90° Celsius (-130° Fahrenheit).

Most meteors that enter the Earth's atmosphere don't make it through the mesosphere. Because they are traveling at extreme speeds, they will break apart and burn up before reaching the Earth's surface.

The next layer above the mesosphere is the thermosphere. The top of the thermosphere can reach altitudes ranging from about 500-1,000 kilometers (310-620 miles). This varying range in height is caused by temperature changes resulting from heat coming from the Sun. The amount of energy released by the Sun varies, and when there is more energy released, temperatures rise, causing the thermosphere to expand. With less energy released, temperatures drop, and the thermosphere contracts, getting smaller. Temperatures also vary significantly between day and night when the Sun is either shining or not shining on a part of the Earth.

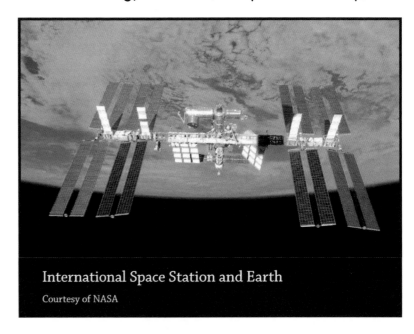

International Space Station and Earth
Courtesy of NASA

The International Space Station orbits within the thermosphere, and auroras (northern and southern lights) occur in this layer.

The exosphere is the outermost layer of the atmosphere and is sometimes considered to be part of space. In this layer, atoms and molecules escape into space freely.

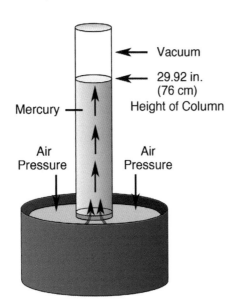

17.4 Atmospheric Pressure

Atmospheric pressure (air pressure) is the force exerted on us by the weight of air molecules in the Earth's atmosphere. It is defined as the force per unit area exerted against a surface by the weight of the air above that surface.

Scientists measure atmospheric pressure with barometers, which are instruments that record the air pressure in millibars (mb) or in inches of mercury. The standard air pressure at sea level is 1,013.25 millibars (29.92 inches of mercury), which is the equivalent of 6.7 kilograms (14.7 pounds).

At sea level the air is warm, there is plenty of oxygen to breathe, and the atmospheric pressure is high. However, if you were to climb a mountain and go up in elevation, away from sea level, you would find that the temperature starts to cool and that there is less oxygen to breathe and less atmospheric pressure. If you were to climb a really high mountain at very high altitudes, you would find that the temperature is even colder, there is even less oxygen to breathe, and much less atmospheric pressure.

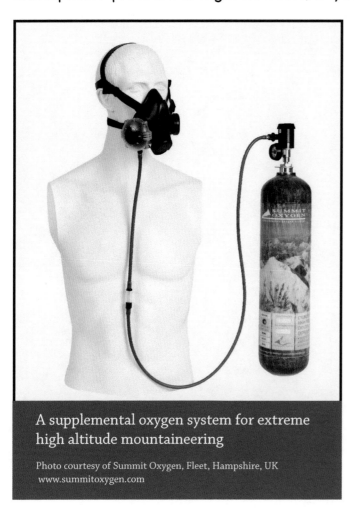

A supplemental oxygen system for extreme high altitude mountaineering

Photo courtesy of Summit Oxygen, Fleet, Hampshire, UK
www.summitoxygen.com

Why does this happen? As you go from sea level to higher and higher elevations, the density of atmospheric gases (the number of gas molecules per volume) decreases, resulting in less air pressure. With less air pressure and fewer gas molecules, the remaining molecules will expand into the available space. As it turns out, when gases expand, they cool. This is why mountain climbers must carry lots of warm clothing and sometimes tanks of oxygen when they climb high mountains.

Mount Everest is the highest place on Earth with an elevation of 8,848 meters (29,029 feet). At the top of the mountain the average air pressure is about 300 millibars. This means that there is only about one-third of the oxygen that there is at sea level.

17.5 Gravity and the Atmosphere

Several factors keep Earth's atmosphere from floating away into space. These factors include the size, or mass, of the gas atoms and molecules, gravity, and the escape velocity. The escape velocity is the minimum speed needed for an object to escape from the gravitational field of a star (sun), planet, or moon. Earth's escape velocity prevents atoms and molecules in the atmosphere from overcoming the force of gravity and escaping into space.

Some other celestial bodies in our solar system do not have atmospheres. The Moon, for example, has such a small atmosphere that it is considered negligible. This is due to the fact that the Moon is less massive than Earth and therefore has a weaker gravitational force. This weaker gravitational force creates a smaller escape velocity for the Moon. Earth's escape velocity is 11 kilometers per second.

Dawn spacecraft is launched to attain escape velocity

Courtesy of NASA/Sandra Joseph, Rafael Hernandez

The Moon's escape velocity is 2.4 kilometers per second. Therefore, it is much easier to escape into space from the Moon than from Earth. Although there is gravity on the Moon, it is too weak to hold a significant atmosphere.

17.6 The Greenhouse Effect

Greenhouses are enclosed structures made of glass or plastic. The glass or plastic lets sunlight enter and then traps some of the heat from the Sun, holding it inside the greenhouse. Trapping the heat keeps a greenhouse warm inside even during winter, allowing plants to grow year round. Our planet's atmosphere traps energy from the Sun in a similar way, and this is called the greenhouse effect. The greenhouse effect occurs because energy from the Sun enters Earth's atmosphere, but not all of it escapes back into space. Some gases act like a greenhouse by absorbing heat and keeping Earth warm. These gases include carbon dioxide, water vapor, and methane, among others.

During the day, sunlight warms the planet. At night the surface of the Earth cools and releases heat back into the atmosphere. Some of this heat does not escape into space but is trapped by greenhouse gases. If this process remains balanced, the Earth will stay at about the same yearly average temperature. If the greenhouse effect is too strong, it won't allow as much heat to escape into space, and then Earth begins to get warmer.

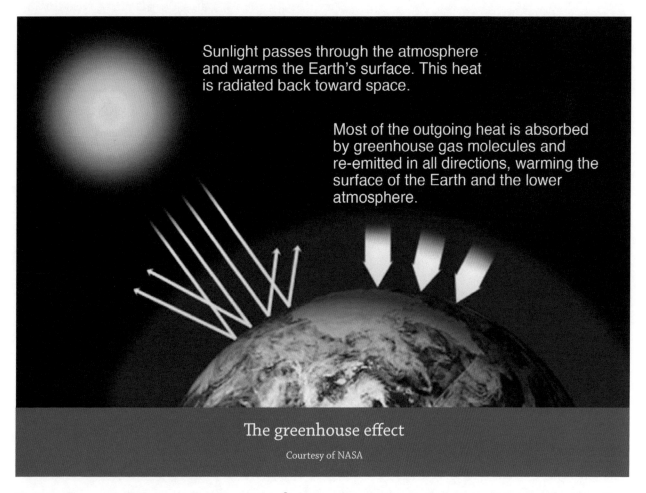

The greenhouse effect
Courtesy of NASA

Every day scientists collect a lot of data about air and ocean temperatures all over the world. The data show that currently there is a rise in the average temperature of the Earth's atmosphere and oceans. Some indications of a warming of Earth are record high temperatures at different locations, the melting of glaciers, severe flooding, droughts, and fires. Although the warming

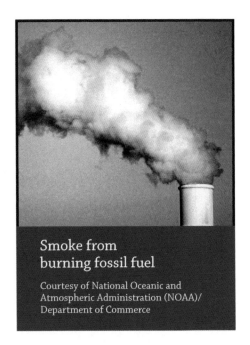

Smoke from burning fossil fuel

Courtesy of National Oceanic and Atmospheric Administration (NOAA)/Department of Commerce

trend is clear, the causes of the warming are under debate.

Measurements show that there is an increase of greenhouse gases in the atmosphere. Many scientists believe that human activity is causing Earth's warming trend, with the increase in greenhouse gases coming mostly from the burning of fossil fuels. These fossil fuels, such as oil, coal, and natural gas, are used for many things, including generating electricity, heating buildings, cooking, and running machinery. This results in a large amount of carbon dioxide being put into the atmosphere by human activities.

Some scientists also theorize that deforestation, or the cutting down or burning of large areas of trees, has increased the amount of carbon dioxide in the atmosphere because there are fewer trees to change carbon dioxide to oxygen through photosynthesis, and the burning itself releases carbon dioxide. Many countries are now working on ways to reduce the burning of fossil fuels and slow down deforestation.

Deforestation   Courtesy of NASA/LBO-ECO Project

Other scientists think that Earth is in a natural warming cycle, while still others think that there are a variety of factors involved. Scientists often have different theories about the same thing, and the warming of Earth is one area of study in which there are different ideas. The functioning of Earth's spheres and the way they work together is so complicated that there is no easy answer to the question of why Earth's climate is changing.

17.7 Summary

- The atmosphere is a thin layer of gases that surrounds the Earth.

- The Earth's atmosphere is divided into five layers: the troposphere, stratosphere, mesosphere, thermosphere, and exosphere.

- The most abundant gas in the atmosphere is nitrogen, with oxygen being the next most abundant.

- Atmospheric pressure is the force exerted on us by the weight of air molecules in the Earth's atmosphere.

- The greenhouse effect is the trapping of energy from the Sun by the Earth's atmosphere. This process keeps the Earth warm.

Chapter 18 Technology in Astronomy

18.1	Introduction	185
18.2	Telescopes	186
18.3	Space Telescopes and Other Satellites	188
18.4	Other Space Tools	190
18.5	Summary	193

Astronomy

Image credits
Background: Orion Nebula viewed by Hubble Telescope
Courtesy of NASA, ESA, T. Megeath
(University of Toledo) and M. Robberto (STScI)
Hubble Telescope Image Courtesy of NASA

18.1 Introduction

In this chapter we will take a look at the tools astronomers use to explore the skies. Tools are an essential part of scientific investigation, and for many centuries astronomers have been using tools to gain a better understanding of the universe.

Even before the invention of the telescope, early astronomers used tools to study the sky. In the 1500s BCE, ancient civilizations used tools to track the movement of the Sun. Stonehenge, a group of huge stones set in a circular shape outside of Amesbury, England is believed to be a kind of solar tracking system. As the Sun moves over the large structure, the shadows mark the summer and winter solstices.

Stonehenge, near Amesbury, England

Today, modern astronomy tools help scientists get accurate readings of star and planetary movements and help them observe stellar objects that they can't see with their eyes. In this chapter we will learn about some of the tools modern astronomers use.

18.2 Telescopes

When Galileo decided to look at the night sky, he used a telescope. The word telescope comes from the Greek prefix *tele-* which means "from afar" or "far off" and *skopein* which means to "see," "watch," or "view." A telescope is an instrument used to see, watch, or view things that are far away.

Galileo is sometimes credited with the invention of the telescope, but in 1608 the Dutch lens maker Hans Lippershey filed the first patent for what would eventually become the telescope. Galileo made many improvements to the Dutch "perspective lens" and was able to greatly increase the magnification. With his powerful lenses, Galileo was able to see that Jupiter has moons!

There are essentially three types of telescopes: refractor telescopes, reflector telescopes, and compound telescopes.

The first telescopes built were refractor telescopes. Refractor telescopes can be found in hobby and toy stores and are the type of telescopes used for rifles.

A refractor telescope houses a lens at one end and an eyepiece at the opposite end of a narrow tube. Light enters one end of the tube and is bent by the lens. The observer looks through the tube from the opposite end and sees the magnified object at the focal point, the spot at which the rays of light entering the lens come together to produce the image.

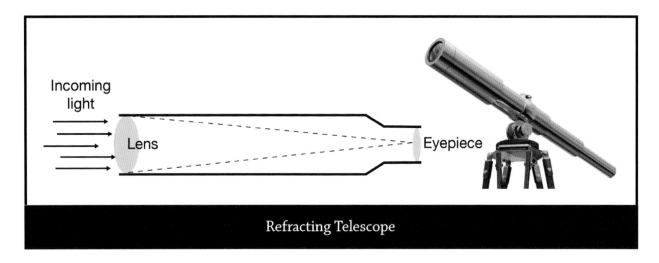

Refracting Telescope

The largest refracting telescope ever constructed was at the Great Paris Exhibition in 1900. It had a focal length (the distance from a lens to its focal point) of 57 meters (187 feet) but was later dismantled after the company went bankrupt. Today the largest refracting telescope in use is housed at the Yerkes Observatory in Chicago.

Yerkes Observatory, Chicago, IL, USA

Reflector telescopes and compound telescopes use a combination of mirrors and lenses to focus incoming light. These telescopes are more complicated than the refractor telescope and can provide better quality images. A common reflector telescope is the Newtonian telescope named after its inventor, Isaac Newton. A Newtonian telescope has a simple design with two mirrors and an eyepiece. Light enters the telescope at the far end, is reflected back by one mirror, and hits a second mirror where it exits to the eyepiece.

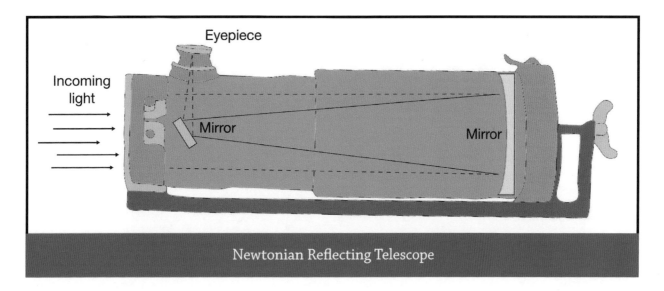
Newtonian Reflecting Telescope

18.3 Space Telescopes and Other Satellites

Although telescopes can be built for viewing planets and stars that are millions of miles away, Earth's atmosphere causes problems for astronomers making observations from the Earth's surface. Light coming from a distant star must pass through Earth's atmosphere before it can be collected by a telescope on the surface. As the light passes through the atmosphere, it can be reflected by tiny atmospheric particles. Atmospheric turbulence causes these particles to move, creating small changes in the optical properties of the air. Before it enters a telescope, the light that is being collected gets bounced around, which makes it appear that the image of the object being viewed is moving. This also makes stars appear to "twinkle." Although twinkling stars are fun to watch, they prevent astronomers from collecting the kind of detailed data that is needed for study.

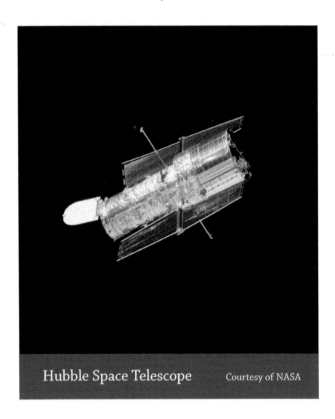
Hubble Space Telescope — Courtesy of NASA

A great way to solve this problem is to set up a telescope outside the Earth's atmosphere. The Hubble Space Telescope is one such telescope. It was placed into orbit around Earth by the Space Shuttle Discovery in 1990. The Hubble Space Telescope is able to take sharp and detailed images of many far distant objects with very little distortion. Also, space telescopes are able to view the universe in wavelengths of light that are blocked by Earth's atmosphere, such as X-rays and gamma rays, and also rays that are partially blocked by the atmosphere, such as ultraviolet and infrared rays.

Our understanding of the universe has been greatly expanded as a result of images taken by the Hubble Space Telescope. Astronomers have been able to learn more about how stars and galaxies form, old stars explode, and galaxies

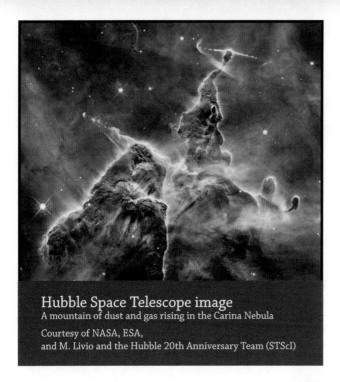

Hubble Space Telescope image
A mountain of dust and gas rising in the Carina Nebula
Courtesy of NASA, ESA,
and M. Livio and the Hubble 20th Anniversary Team (STScI)

collide. Much has also been learned about black holes, nebulae, and planets that are orbiting other stars.

Another satellite is the Kepler Space Telescope which was placed into orbit in 2009 for the purpose of looking for planets outside our solar system. Even though Kepler explores a very small portion of the sky, it has discovered many planets, and scientists now think that most stars have planets orbiting them. We'll take another look at the Kepler Space Telescope in Chapter 21.

The largest and most complex satellite is the International Space Station (ISS). The space agencies most involved in constructing and operating the space station are from the United States, Russia, Europe, Japan, and Canada. The first space station module was launched in 1998 and more modules have been added over the years. The first crew arrived at the ISS in 2000, and it is now continuously occupied, with crews from different countries coming and going. Research done on the ISS includes studies of human health and life sciences, testing technologies that may be used in future space explorations, and research in various fields such as physical sciences and earth and space science.

A meeting of the minds aboard the International Space Station on 7 March 2015

Members of Expedition 42: astronaut US, Barry Wilmore (Commander) Top, Upside down, to the right Russian cosmonaut Elena Serova, & ESA astronaut Samantha Cristoforetti. Bottom center US astronaut Terry Virts, top left cosmonauts Alexander Samokutyaev and Anton Shkaplerov.
Courtesy of NASA

There are currently over 1000 operational satellites orbiting Earth. Satellites have become an important part of our lives, with uses for television, cell phones, GPS, weather forecasting, and monitoring changes to Earth's environment and climate. They also have many other scientific uses such as the study of the Sun, Moon, and magnetosphere. Satellites have greatly increased our knowledge of Earth and the cosmos and make it much easier for us to communicate.

18.4 Other Space Tools

Space probes, landers, and rovers are other tools scientists use to explore space.

A space probe is a robotic spaceship that can travel far distances, capturing images and collecting data. Voyager 1 and 2 are space probes that were launched in 1977 to explore Jupiter and Saturn. Voyager 1 has continued on its journey to become the first spacecraft to leave our solar system and is now traveling in interstellar space. After passing Jupiter and Saturn, Voyager 2 also went past Uranus and Neptune, sending data back to Earth, and it is now nearing interstellar space. Both Voyager 1 and 2 are still sending signals back to Earth for scientific analysis.

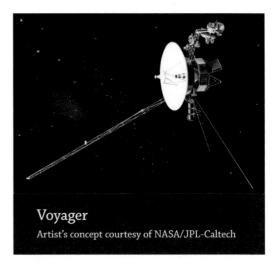
Voyager
Artist's concept courtesy of NASA/JPL-Caltech

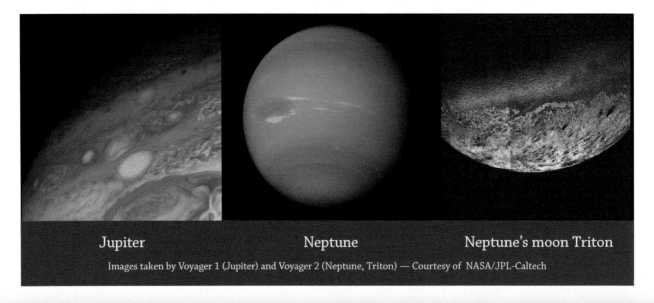
Jupiter   Neptune   Neptune's moon Triton
Images taken by Voyager 1 (Jupiter) and Voyager 2 (Neptune, Triton) — Courtesy of NASA/JPL-Caltech

Spacecraft that go into orbit around celestial bodies other than Earth are called orbiters. For example, NASA's Mars Atmosphere and Volatile Evolution (MAVEN, launched in 2013) is studying the atmosphere of Mars, and India's Mars Orbiter Mission (MOM, launched in 2013) is looking at the surface features and mineralogy as well as the atmosphere of Mars. Both of these orbiter missions have a fairly short life expectancy—depending on when they run out of fuel.

MAVEN orbiter observes Mars aurora
Artist's concept courtesy of University of Colorado and NASA

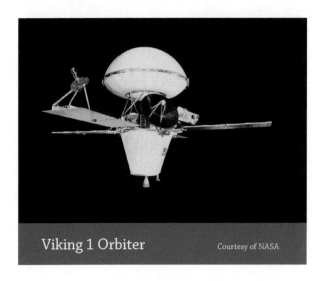

Viking 1 Orbiter Courtesy of NASA

Like space probes and orbiters, a lander is a robotic spacecraft, but it is able to land on the surface of planets, asteroids (small celestial bodies made mostly of rock and minerals), or comets (large chunks of ice and dirt). NASA's Viking 1 was a spacecraft that consisted of both an orbiter and a lander, and the Viking 1 lander was the first to successfully land on the surface of Mars and send images back to Earth. Viking 1 touched down on Mars in July 1976 and continued collecting data for six years. The Viking 1 orbiter sent back images of Mars from space from 1976 through 1980.

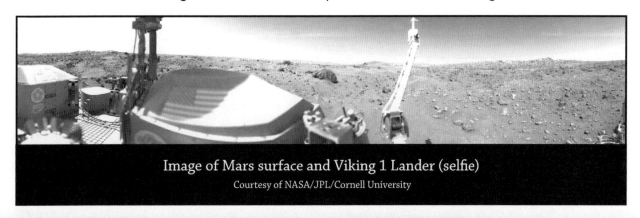

Image of Mars surface and Viking 1 Lander (selfie)
Courtesy of NASA/JPL/Cornell University

A more recent orbiter-lander combination is the Cassini-Huygens mission which is a cooperative project of NASA, the European Space Agency, and the Italian Space Agency. In 2004 the Cassini spacecraft released the Huygens lander over Saturn's largest moon, Titan. Huygens had a safe landing and sent back data that revealed many interesting and unexpected details about Titan. Cassini has gone on to explore Saturn and its moons and has also found many unexpected features, such as moons covered with ice.

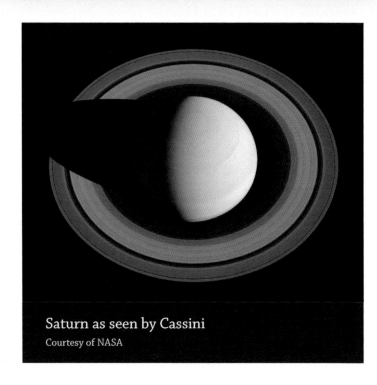

Saturn as seen by Cassini
Courtesy of NASA

Mars rover Curiosity (selfie)
Courtesy of NASA/JPL-Caltech/Malin Space Science Systems

A rover is a lander that can move. A rover is an automated machine that can travel across a planetary surface, gathering data as it goes. Since the mid 1990s several NASA rovers have landed on Mars. The first was Sojourner in 1997 followed by Spirit and Opportunity in 2004. Although Opportunity continues to collect data, Spirit became stuck in the sand, and in March 2010, it stopped transmitting data.

The most recent Mars rover is Curiosity which was launched in November 2011. Curiosity's mission includes: looking for the chemical building blocks of life, such as organic carbon compounds; investigating the chemical composition of the Martian surface; interpreting the ways rocks and soils may have formed; and looking for the cycling of carbon dioxide and water. The rovers are helping scientists learn about the makeup of Mars which will aid in planning for astronauts to successfully land on Mars.

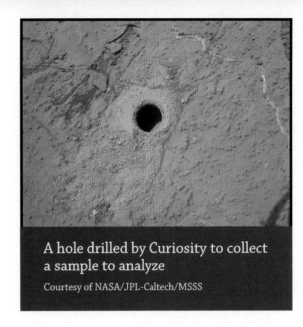

A hole drilled by Curiosity to collect a sample to analyze
Courtesy of NASA/JPL-Caltech/MSSS

These are just a very few examples of the many different spacecraft that have been launched by different countries. Space exploration continues to expand rapidly, with each discovery building on the last and increasing our knowledge of what lies beyond Earth.

18.5 Summary

- Astronomers use tools to explore the cosmos.
- Telescopes are used to magnify faraway objects.
- Space telescopes avoid atmospheric distortion.
- Satellites increase our knowledge of Earth and the cosmos and have become an important part of our lives.
- Modern astronomers can utilize space tools to collect data from far distant locations. Space probes, orbiters, landers, and rovers are among the tools used.

Chapter 19 Time, Clocks, and the Stars

19.1	Introduction	195
19.2	Reading a Star Atlas	197
19.3	Time	199
19.4	Celestial Clocks	202
19.5	Summary	204

Astronomy

19.1 Introduction

What would you do if you were on a hike with a group of friends and suddenly found yourself separated? Imagine that your compass needle is stuck and as you take your whistle from your pocket to call your friends, it falls out of sight in a crack between two big rocks that are too heavy to lift or move. The day is coming to a close and the night stars are beginning to shine. If you don't know how to navigate with the stars, the best thing to do would be to sit down and hope that your friends or someone else will find you. However, if you do know about stars, the constellations, and how to interpret their position in the night sky, you could find your way back to the group camp or find your way home.

In ancient times people looked to the stars for inspiration, religious meaning, and navigation. One way early people gave meaning to the stars was to look for patterns, create groups of stars, and give them names. These named groups of stars are called constellations.

We don't know the exact date the first constellations were named, but ancient Egyptians, Sumerians, and Chaldeans are believed to have known many of our present-day constellations. It appears that by 2000 BCE most of the main constellations in the Northern Hemisphere had been recorded. The Greeks and Romans took over where their ancient ancestors left off, using Greek and Latin to name many of the constellations after heroes, animals, and mythical objects.

Mapping the stars is called celestial cartography or uranography. Cartography is the art and science of making maps. Uranography comes from the Greek word *uranos* which means "heavens" and *graphe*, which means "to write," so uranography means the "writing of the heavens." Although constellations are visible from both the Northern and Southern Hemispheres, it wasn't until after the 15th century CE that constellations in the Southern Hemisphere were recorded by European explorers.

There are forty-eight original constellations which include the constellations known to ancient Greek, Roman, and western Asian people. Today, the IAU (International Astronomical Union) recognizes 88 constellations. The newer constellations are found mostly in the Southern Hemisphere and were charted by Europeans as they explored that part of the globe. However, we now know that ancient cultures all over the world recognized constellations.

Mapping the stars was a popular activity for many early astronomers, and with the invention of the telescope, modern astronomers began to focus primarily on determining the accurate position of stars and celestial objects rather than using constellations for reference. In 1875 the German astronomer Friedrich Argelander (1799-1875 CE) published a catalog of the locations of 325,000 stars. This star catalog, called the Bonner Durchmusterung, was simply a grid that showed the positions and magnitudes of stars without relating them to the constellations. This star catalog is still being used in revised and updated forms. Another star catalog still in use today is Norton's Star Atlas which was first published in 1910 and was based on a star catalog created by Belgian astronomer Jean-Charles Houzeau. Norton's Star Atlas has also been revised and updated many times.

From 1989-1993 the Hipparcos satellite gathered data that was used to create the Hipparcos Star Catalog which accurately mapped over 100,000 of the brightest stars and the Tycho Star Catalog which mapped over 2 million dimmer stars with slightly less accuracy. The Hubble Space Telescope has also been used to catalog stars.

Astronomy—Chapter 19: Time, Clocks, and the Stars 197

Today, there are different star catalogs, or star atlases. In addition to the Hubble and Hipparcos/Tycho star catalogs, there are a number of print and computer generated star atlases that map not only the stars in our galaxy, but deep space stars, nebulae, and celestial objects. Many star maps include the constellations in addition to the positions, magnitudes (brightness), and movement of individual stars.

19.2 Reading a Star Atlas

A star atlas or star map can be quite overwhelming for the beginning astronomer. Modern star atlases map over 450,000 stars, planets, and other celestial objects. Just locating where you are relative to the stars in the sky can be daunting. Not only are there thousands of tiny dots representing the locations and brightness of the stars, but the stars' locations in the sky are constantly changing with the seasons.

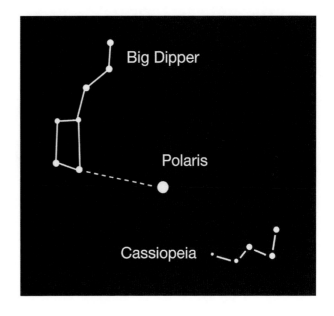

For this reason, it's easiest to orient yourself starting with familiar landmarks like the constellations and asterisms, which are groups of stars that are smaller than constellations and may be part of a constellation. Patterns are easier to see than a single star in a cluster of stars, and by locating a constellation or asterism in the sky you can find the surrounding stars mapped in the atlas. For example, if you are in the Northern Hemisphere, one of the easiest asterisms to locate is the Big Dipper. Recall that the two stars in the part of the bowl of the Big Dipper that are farthest from the handle point to Polaris, the North Star. When you are facing Polaris, you are facing north. Polaris is in between the Big Dipper and the constellation called Cassiopeia. Cassiopeia is a W-shaped set of stars that is easy to find.

198 Exploring the Building Blocks of Science: Book 6

Once you find the Big Dipper, Polaris, and Cassiopeia, you can use the star atlas as a kind of road map. By holding the star map above your head, and turning it to align with the constellation landmarks, you will be able to identify stars that are outside the constellation. Star maps vary with a given day or season since the stars change position throughout the year and some are only visible at certain times of the year.

Star atlases also come as calendar charts. Each calendar chart lists the stars and constellations that are visible from different locations on Earth and shows the positions of the stars for a particular month. Some star calendars show thousands of stars, but many star calendars only show a few hundred stars, making it easier to navigate.

Star atlases not only map the location of the stars but indicate a star's brightness. Brighter stars are shown as large dots, with less bright stars shown as smaller dots. Star atlases also map galaxies, clusters, and nebulae using different symbols such as dotted circles, closed circles, and ovals.

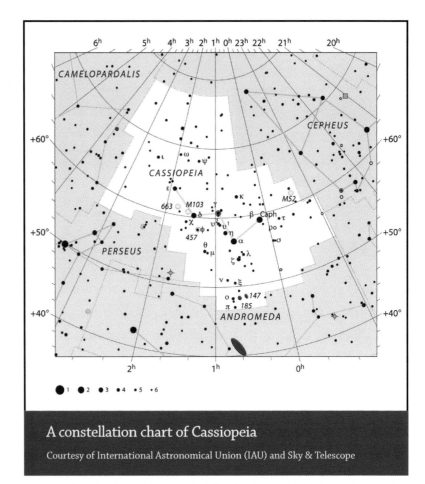

A constellation chart of Cassiopeia
Courtesy of International Astronomical Union (IAU) and Sky & Telescope

19.3 Time

What time is it? How do you determine the time when you need to get up in the morning, go to class during the day, or go to sleep at night? If you are like most people in the modern world, you probably use your wristwatch, wall clock, or digital time on your phone. But what is time exactly and how do you know your watch is correct?

An example of a sundial

In the morning you can see the Sun rise and in the evening you can see the Sun set. The Sun comes up over the eastern horizon and sets in the western horizon after spending a certain amount of time in the sky illuminating your day. The time it takes for the Sun to go from its highest position in the sky on one day to its highest position on the next day is called a solar day. If you use a sundial, you can measure when the Sun is at its highest position in the sky and how long it takes the Sun to go from one position in the sky to the next. A sundial records the Sun's daily motion across the sky and gives apparent solar time. Apparent solar time is time measured from the direct observation of the Sun. But on cloudy days and at night a sundial won't work. Although apparent solar time is a natural way to keep track of time, it isn't accurate enough for our modern world, so other ways of keeping time have been devised.

If you live anywhere on Earth other than near the equator, you will notice that due to the tilt of Earth on its axis the length of a day changes with the seasons. In the Northern and Southern Hemispheres during the winter months the days are shorter than in the summer months. At the equator the length of a day stays about the same all year, and the farther you are from the equator, the more variation there is in the length of the day throughout the seasons. Also, because Earth's orbit is slightly elliptical, Earth's distance from the Sun varies, with Earth being closer to the Sun in the fall and early winter months. This causes the length of day in apparent solar time to vary by as much as 15 minutes between the Earth's closest position and farthest position from the Sun.

To correct for the length of the day changing over the course of a year, astronomers use mean solar time. Mean solar time is based on apparent solar time averaged over the course of a year. In other words, if you measure the length of each of the apparent solar days in a year, add together the day lengths for all the days in the year, and then divide this by the number of days in the year, the result is the average length of a solar day, which is mean solar time. A standard watch uses mean solar time that is divided into hours, minutes, and seconds. Global time zones are created using mean solar time.

Mean solar time tells us that it takes 24 hours for Earth to spin once around its axis, which is the length of one mean solar day. But at the same time that Earth is spinning on its axis, it is also rotating around the Sun. Earth travels so fast in its orbit around the Sun that by the time Earth has made one rotation on its axis, it has traveled 2.5 million kilometers in its orbit around the Sun! So, in one mean solar day, Earth has not only rotated once on its axis but has also changed its position in space relative to the Sun.

For example, let's say you are in New York City at noon and you observe that the Sun is at its highest position in the sky. The next day you look at your watch and it is 24 hours later with the Sun once again in its highest position in the sky. On each of these days at noon in mean solar time, the Sun is directly above New York City (the Sun is in its highest position in the sky). The problem is that the Earth, as well as rotating once on its axis, has also moved 2.5 million kilometers in its orbit around the Sun from one noon to the next. Although due to Earth's rotation the Sun is directly above New York City at two noontimes that are 24 hours apart in mean solar time, the *actual* rotation of the Earth (or true rotation) has only taken 23 hours and 56 minutes. Why is this? It's the change of the position of Earth relative to the Sun that makes the length of the mean solar day longer than the length of time for one true rotation. This results in a difference of 4 minutes between how long it takes Earth to make one true rotation on its axis and the length of one day in mean solar time. The length of time it takes for Earth to make one true rotation is called a sidereal day.

You can observe the difference between the length of time of one mean solar day and the length of one sidereal day (one true rotation) if you observe the same stars at the same location in the sky on several consecutive nights. When the stars arrive at the location you've noted, you will discover that the time on your watch will be about four minutes earlier each night. This four minute difference is the result of the difference between mean solar time and the actual rotation time of Earth.

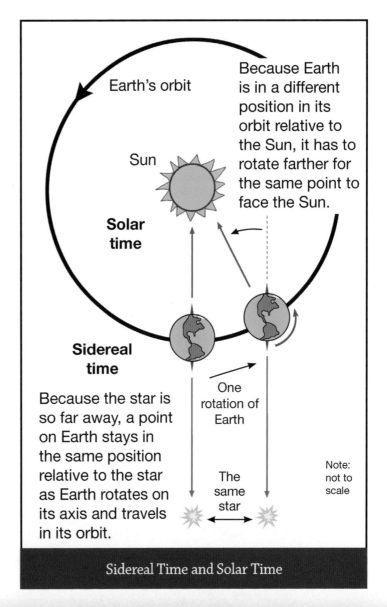

Sidereal Time and Solar Time

Astronomers developed sidereal time, or star time, to measure time more accurately by using Earth's position relative to distant stars rather than relative to the Sun. The term sidereal comes from the Latin word *sidus* which means "star." Using sidereal time, astronomers are able to calculate the true rotation of Earth. Similar to how a mean solar day is measured by the

Sun's highest position in the sky from one day to the next, a sidereal day is measured from the time a distant star appears in its highest position from one night to the next. Because the stars are so far away, Earth's orbit does not affect the position of where the stars appear in the sky—only the rotation of Earth does.

Although sidereal time is a more accurate measure of Earth's rotation on its axis, it is most useful to astronomers. For everyday use, solar time works best.

19.4 Celestial Clocks

In the modern world our days are divided up into segments of hours, minutes, and seconds. Because we no longer rely on observing the location of the Sun to tell time, we use clocks to help us make sure we get to our appointments on time, pick up the laundry on time, eat breakfast, lunch and dinner on time and go to bed on time.

Our standard clocks don't take into account the movement of the Moon, other planets, or the stars. A celestial clock, or astronomical clock, on the other hand, is an instrument first built by ancient scholars to provide information about astronomical movements of celestial bodies as well as keeping track of time.

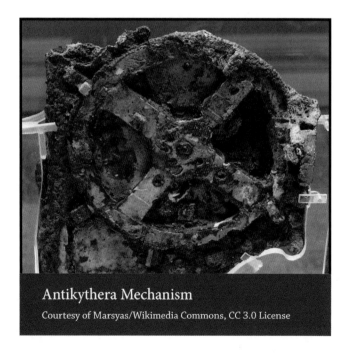

Antikythera Mechanism
Courtesy of Marsyas/Wikimedia Commons, CC 3.0 License

The first astronomical clock we know of is the Antikythera Mechanism built 2000 years ago by the ancient Greeks. Pieces of the mechanism were discovered by divers in 1900 near the tiny island Antikythera in Greece. Using x-rays to peer into the body of the device and computers to reconstruct how the device worked, scientists have suggested that it was an incredibly accurate astronomical clock able to replicate the irregular motions of the Moon, track the position of the

Earth in its orbit around the Sun, and determine the position of Venus, Mars, Jupiter, and Saturn for any chosen date.

Another early astronomical clock was designed by Su Sung of China, built in 1092, and ran until 1126 when the Sung Dynasty was overtaken. This clock is called the Cosmic Engine and is an astronomical clock powered by falling water or falling mercury. The original clock tower was 30 feet tall with a series of interlocking gears rotating with precision. The clock displayed the positions of the Sun, Moon, and planets.

Su Sung's astronomical clock

Prague astronomical clock
Courtesy of Andrew Shiva

Astronomical clocks became a spectacle in the European world during the Middle Ages. One of the most famous old astronomical clocks still existing is located in the town square in Prague, the capital of the Czech Republic. The clock was built by Mikulas of Kadan in 1410 and consists of an astronomical dial, a calendar dial, and a window with rotating characters.

19.5 Summary

- Mapping the stars is called celestial cartography or uranography.

- Apparent solar time is the time it takes for the Earth to complete one rotation around its axis (complete one day) based on the position of the Sun.

- Mean solar time is the average of apparent solar time over a full year.

- Sidereal time is the actual measurement of Earth's rotation around its axis based on the position of a fixed star.

- Celestial clocks record time plus movements of celestial bodies.

- A star atlas is like a road map of the night sky, mapping the locations and brightness of stars, planets, and other celestial bodies.

Chapter 20 Our Solar System

20.1	Introduction	206
20.2	Planetary Position	206
20.3	Planetary Orbits	207
20.4	Asteroids, Meteorites, and Comets	209
20.5	Habitable Earth	212
20.6	Summary	213

Astronomy

20.1 Introduction

In a previous book we examined the eight planets of our solar system. We saw that the planets are divided into two broad categories: terrestrial planets and Jovian planets. We discovered that the four planets closest to the Sun are terrestrial planets made mostly of rock, like Earth, and the four outer planets are Jovian planets made mostly of gases, like Jupiter.

In this chapter we will take a closer look at our solar system. A solar system is a group of celestial bodies and the one or more suns they orbit. Our solar system has eight planets orbiting a single sun.

20.2 Planetary Position

If we look at our entire system of planets, we see that the Sun is in the center of the solar system with the planets orbiting the Sun in a particular order. Mercury orbits closest to the Sun followed by Venus, Earth, Mars, Jupiter, Saturn, Uranus, and finally Neptune.

Because the distance from the Sun to the planets is very large, astronomers measure planetary distances in units called astronomical units, or AU. One AU is equal to 149,597,870.7 kilometers (92,955,801 miles). To get an idea of just how far one AU is, imagine that you had to drive from the Earth to the Sun (1 AU) in your car going 97 kilometers per hour

(60 miles per hour). To get to the Sun this way, it would take 1,549,263 hours or 64,552 days, or about 177 years!

Using AU to measure the distance of the planets from the Sun, you can see that the four terrestrial planets are relatively close together. All of the terrestrial planets are less than 2 AU from the Sun, with Mercury the closest at 0.387 AU and Mars the farthest at 1.524 AU.

Planets and their distances from the Sun — Courtesy of NASA

In between the terrestrial planets and the Jovian planets is a huge 4 AU space. Jupiter, the closest of the Jovian planets, is 5.2 AU from the Sun, and Neptune, the farthest of the Jovian planets, is an incredibly far 30 AU from the Sun!

20.3 Planetary Orbits

An orbit is defined as the gravitational curved path of one celestial body moving around another celestial body. In other words, the orbit is the "road" a planet travels as it circles the Sun, and the Sun's gravity is what holds the planet in its orbit.

All of the planets orbit the Sun in a counterclockwise direction, and if we take a look straight down at the planetary orbits, we discover that the orbits look almost circular. They are not fully circular and so are technically elliptical, but they are not as elliptical as many people think they are.

One common misconception about Earth's seasons is that it is Earth's orbit that gives us the summer and winter months. However, by examining Earth's orbit it's easy to see that the difference between Earth's farthest and closest distance from the Sun is very small. In other words, as Earth orbits the Sun, Earth's closest position to the Sun is not significantly different from its farthest position from the Sun. The seasons are determined by Earth's tilt on its axis, not its distance from the Sun. One pole of the Earth is tilted toward the Sun in the summer months and away from the Sun in the winter months.

Because there is such a large gap between Mars and Jupiter, astronomers place the planets in two groups. The terrestrial planets make up the inner solar system and the Jovian planets make up the outer solar system.

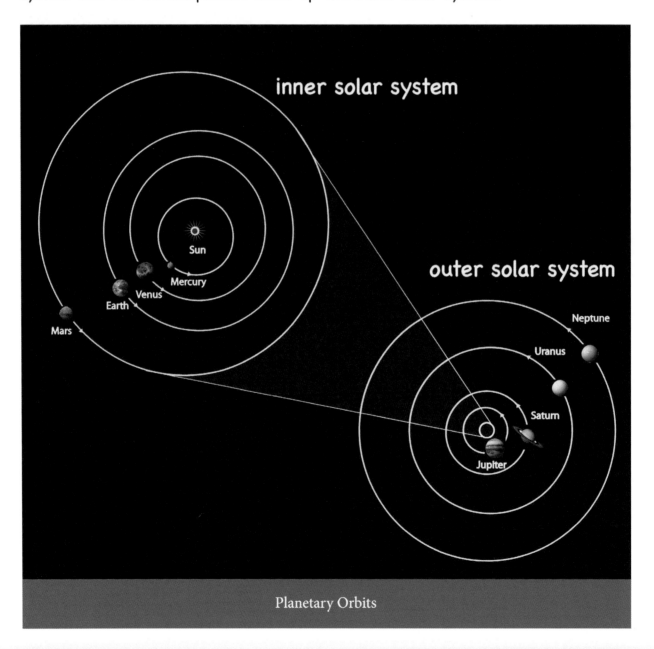

Planetary Orbits

20.4 Asteroids, Meteorites, and Comets

The gap between Mars and Jupiter is not empty space but instead is home to millions of asteroids. The word asteroid comes from the Greek word *aster* which means "star." An asteroid is a small celestial body made mostly of rock and minerals, but when an asteroid is viewed in the sky, it can resemble a small star. However, asteroids are not real stars like our Sun because they are only reflecting light from the Sun rather than emitting their own light. The asteroids between Mars and Jupiter occupy an area known as the Asteroid Belt. Asteroids also exist outside the Asteroid Belt.

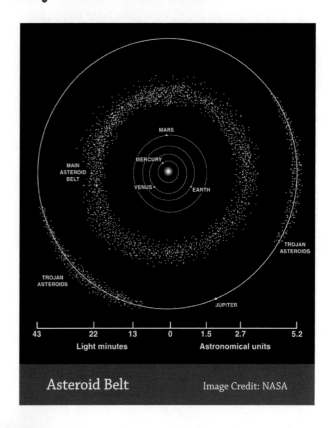

Asteroid Belt Image Credit: NASA

Asteroid Vesta as seen by NASA's Dawn spacecraft
Courtesy of NASA/JPL-Caltech/UCAL/MPS/DLR/IDA

Scientists estimate that there are 1-2 million asteroids in the Asteroid Belt that are more than 1 km (.62 mi.) in diameter and millions more that are smaller. A few are much larger, like Asteroid Lutetia which is 100 km (62 mi.) in diameter and Asteroid Vesta which is about 525 kilometers (326 mi.) in diameter. Asteroids often have irregular shapes and some have small moons orbiting them.

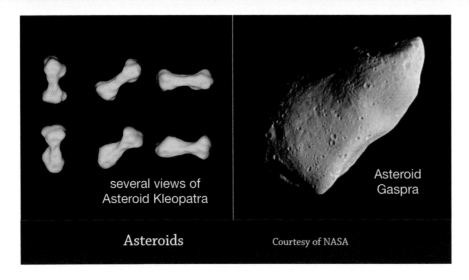

several views of Asteroid Kleopatra

Asteroid Gaspra

Asteroids

Courtesy of NASA

Asteroid Gaspra is an asteroid with an elongated body, and Asteroid Kleopatra has a dog bone shape.

Although there are great distances between asteroids in the Asteroid Belt, asteroids sometimes collide. Because asteroids are moving at great speeds, when they collide, the force of the impact is more than sufficient to shatter rock. Many asteroids have craters on their surface as a result of these high impact collisions.

Asteroids are also found outside the Asteroid Belt and do occasionally impact Earth. Small asteroids, if they cross into the Earth's atmosphere, are called meteors. They often break up into smaller pieces and burn up before reaching the surface of the Earth. Meteors that reach the Earth's surface are called meteorites. Depending on their composition, meteorites are called "stones" or "stony irons."

Meteor over Mauna Kea, Hawaii
(the Big Dipper can be faintly seen over the observatory)

Courtesy of Gemini Observatory/Joy Pollard

Scientists are researching asteroids to find out if they contain materials that could be mined in the future. In 2005 the Japan Aerospace Exploration Agency (JAXA) spacecraft Hayabusa landed on the asteroid Itokawa, and in 2010 Hayabusa brought back to Earth a small sample of asteroid dust for analysis. It appears that rather than being solid rock, Itokawa consists of a group of rocks held together by gravity. There are also missions being planned by several countries to see if it is possible to use a controlled impact to change the orbit of an asteroid, the idea being that if an asteroid is headed toward a collision with Earth, it could be deflected so it would miss Earth.

A comet is another type of celestial body found in our solar system. Comets are large chucks of dirty ice. Some comets have an orbit that brings them close to the Sun. When this happens, the Sun's heat vaporizes some of the ice, changing the frozen water and frozen gases directly from the solid state to the gaseous state and creating long tails of gas and dust particles that are visible when the particles reflect light from the Sun.

Two famous comets that can be easily seen when their orbits bring them close to Earth are Halley's Comet and the Hale-Bopp Comet. In 1986 as Halley's Comet passed close to Earth, several spacecraft were able to get close enough to gather information about it. Halley's Comet has a potato-shaped center about 15 kilometers (9 miles) long and a long tail made of various frozen gases such as carbon dioxide, methane, and ammonia. In 1997 Hale-Bopp Comet passed by Earth, displaying a beautiful fluorescent blue-white tail made of ionized carbon monoxide molecules.

Comets
Courtesy of NASA

In 2004 the European Space Agency (ESA) launched the Rosetta spacecraft whose mission was to orbit Comet 67P/Churyumov-Gerasimenko and send data back to Earth. It took ten years for Rosetta to arrive at the comet, and Rosetta orbited the comet until the mission ended in 2016. Rosetta also released a lander to the comet's surface, but when it landed, it didn't work. Rosetta collected data as the comet's orbit took it closer to the Sun, enabling scientists to observe the comet as it was "activated" by energy from the Sun, causing the frozen gases to begin to vaporize.

Comet 67P/Churyumov-Gerasimenko as seen by ESA's Rosetta spacecraft

Courtesy of European Space Agency (ESA)/Rosetta NAVCAM CC BY-SA IGO 3.0

20.5 Habitable Earth

Within our solar system, as far as we know, there are no other planets, moons, or other celestial bodies that can support life as we know it. Scientists have long been searching for other planets like Earth that could be home to extraterrestrial life—life that exists outside the Earth's system. But so far, science fiction novels are the only place extraterrestrial life exists.

What makes Earth uniquely habitable?

One unique feature of Earth is our atmosphere. Our transparent atmosphere helps maintain the necessary balance of water, gas, and energy. No other atmosphere like Earth's has yet been found to exist.

All known life is dependent on liquid water, and the Earth is located at just the right distance from the Sun for liquid water to exist. A little too close and our oceans would boil, leaving no water for life. A little too far away and Earth and our oceans would freeze and be too cold to support life.

The Moon stabilizes Earth's tilt, and the large planets, Jupiter and Saturn, shield the inner solar system from receiving too many impacts by comets. So both the Moon and the planets help stabilize Earth's habitability.

Scientists are using many different space telescopes, probes, and landers to look for planets outside our solar system that are at the right distance from the Sun to have liquid water and that might have the other conditions necessary for life as we know it. Within our solar system, scientists think they may have discovered liquid water below the ice on a moon of Jupiter called Europa and a moon of Saturn called Enceladus, but it is not yet known if some form of life exists on either moon. Some scientists think that microbes such as archaea might be able to live in the extreme conditions on these moons.

20.6 Summary

- The terrestrial planets (Mercury, Venus, Earth, and Mars) make up the inner solar system and are "close" to the Sun (less than 2 AU).

- The Jovian planets (Jupiter, Saturn, Uranus, and Neptune) make up the outer solar system, and are "far" from the Sun (more than 5 AU from the Sun).

- Each of the eight planets has a slightly elliptical orbit (very close to circular).

- Asteroids exist throughout the solar system, but most are found in the Asteroid Belt between Mars and Jupiter.

- Earth is the only known habitable celestial body in our solar system and is uniquely suited for life.

Chapter 21 Other Solar Systems

21.1	Introduction	215
21.2	Closest Stars	215
21.3	Brightest and Largest Stars	217
21.4	Planets Near Other Stars	218
21.5	The Circumstellar Habitable Zone	220
21.6	Summary	222

Astronomy

21.1 Introduction

In Chapter 20 we explored our solar system. We saw that the Sun is the center of our solar system, and the planets orbit the Sun in a counterclockwise direction. We discovered that Earth is unique among the planets in our solar system in that it is the only planet that we know to support life. But what about planets in other solar systems? Are there other suns in our universe that have planets orbiting them? Do any of these other solar systems support life? In this chapter we will explore some of our neighboring stars.

21.2 Closest Stars

If we look outside our solar system, we discover that there are countless other stars and solar systems. The closest stars to our solar system are actually part of a triple-star system called the Alpha Centauri system.

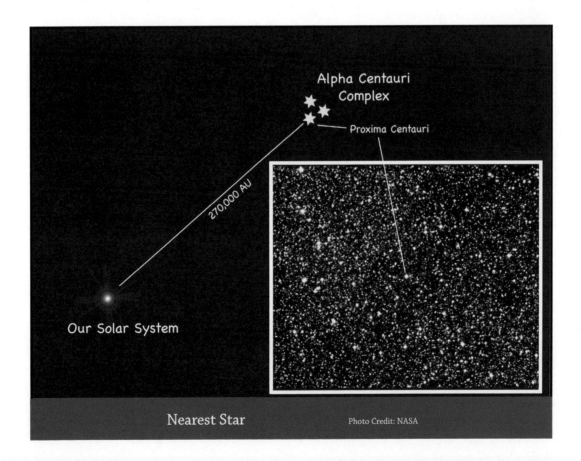

The three stars in this triple-star system are bound together by gravity. Two of these stars, Alpha Centauri A and Alpha Centauri B, are similar to our Sun and orbit each other. The third star, Proxima Centauri, is the star that is closest to Earth, and it orbits Alpha Centauri A and Alpha Centauri B.

Even though Proxima Centauri is closest to our solar system, it is still about 270,000 AU away. This means that the distance from Earth to Proxima Centauri is almost 300,000 times the distance from Earth to our Sun!

The next nearest stars to our solar system are Barnard's Star, Lalande 21185, and Wolf 359. All of these stars are many millions of miles away from our solar system.

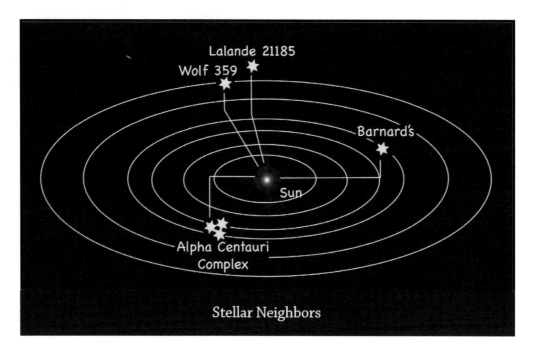

Because stellar distances are extremely large, astronomers measure these distances in parsecs. We saw in Chapter 20 that Earth is 1 AU from the Sun, which is about 150 million kilometers (93 million miles). A parsec is equal to 206,260 AUs or about 19,000,000,000,000 miles! It is easy to see why astronomers measure stellar distances in parsecs.

Barnard's Star is roughly 1.1 parsecs from Earth, Lalande 21185 is 1.4 parsecs from Earth, and Wolf 359 is about 2.4 parsecs from Earth.

21.3 Brightest and Largest Stars

The brightest stars in the sky are not necessarily the closest or largest stars. Sirius is the brightest star in the sky, and it is 2.6 parsecs away from our Sun. It is not as close as the Alpha Centauri star system, but Sirius is 20 times brighter than our Sun and over twice as large.

Sirius can be found in the Canis Major constellation. Sirius has a secondary star associated with it called Sirius B which is significantly dimmer than Sirius.

Sirius
Photo Credit: ROSAT, MPE, NASA
Courtesy of Skyview (http://skyview.gsfc.nasa.gov)

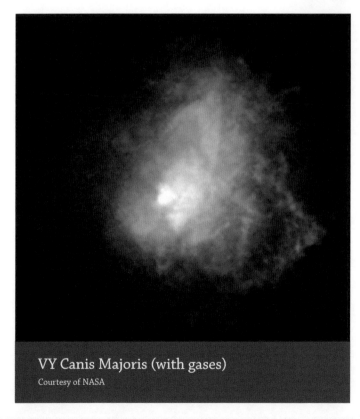
VY Canis Majoris (with gases)
Courtesy of NASA

The largest star visible in the night sky is the star VY Canis Majoris (VY CMa). This star dwarfs our Sun by several magnitudes. A magnitude is the measurement of the brightness of a celestial body.

VY Canis Majoris is considered a red hypergiant star and is located in the constellation Canis Major. VY CMa is very far away from our Sun at a distance of 1500 parsecs. It is a solitary star and does not have multiple stars associated with it. VY CMa is a very active star that emits large amounts of gas during stellar outbursts, which are eruptions of electrically charged particles from a star's surface. These stellar outbursts result in mass being ejected from the star.

21.4 Planets Near Other Stars

Because the large amount of light generated by a star hides the much smaller planets that lie close to the star, it has been difficult to confirm the existence of extrasolar planets. Extrasolar planets (or exoplanets) are planets that orbit stars outside our solar system. Finding exoplanets has been discussed at least since the middle of the 20th century, but it wasn't until 1994 that the existence of planets outside our solar system was confirmed.

Exoplanets as seen by the Hubble Space Telescope (indicated in green)
Courtesy of NASA

Astronomy—Chapter 21: Other Solar Systems

The Kepler Space Telescope
Artist's concept courtesy of NASA/Kepler Mission/Wendy Stenzel

In March 2009 NASA launched the Kepler Space Telescope into an orbit around the Sun with the mission of finding habitable planets around other stars in the Milky Way Galaxy. The Kepler Space Telescope is named after Johannes Kepler (1571-1630), the German astronomer who developed the laws of planetary motion to describe how the planets move around the Sun.

The Kepler Space Telescope uses the transit method to find exoplanets. With this method Kepler detects planets by observing the tiny dimming of a star's brightness as a planet passes in front of, or transits, the star it is orbiting. In order for the transit method to work, the exoplanet's orbit must be aligned with the telescope's line of sight. In other words, Kepler needs to have an edge-on view of the orbit for the telescope to be able see the effect of the planet passing in front of its star. If Kepler is looking at an orbit from the "top," the planet cannot be seen transiting its star. Since a planetary orbit can be oriented at any angle relative to Kepler's point of view, many exoplanets will not be visible to Kepler.

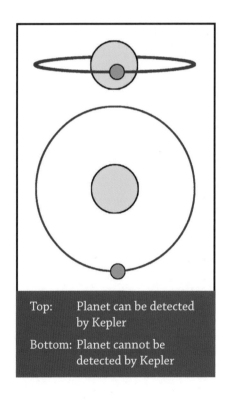

Top: Planet can be detected by Kepler

Bottom: Planet cannot be detected by Kepler

Using data sent to Earth from the Kepler Space Telescope, by November 2016 astronomers had confirmed the existence of over 3400 exoplanets and detected thousands of possible planets. Scientists estimate that in our galaxy alone there may be billions of planets that are of a size similar to that of Earth, and the total number of planets in the Milky Way may be hundreds of billions. It is thought that most of the stars in the Milky Way have planets orbiting them.

GPI image of planetary system HR 8799 showing 3 of its planets

Courtesy of Christian Marois (NRC Canada), Patrick Ingraham (Stanford University), and the GPI Team

At its Gemini South Telescope in Chile, the Gemini Observatory has installed a new instrument called the Gemini Planet Imager (GPI). The GPI is able to find Jupiter-like planets from an Earth-based observatory rather than from a space telescope. It images planets directly by detecting infrared (heat) radiation from gas planets. The Kepler Space Telescope, on the other hand, uses visible light and the transit method to indirectly detect a planet. Each telescope views only a small portion of the galaxy.

Astronomers classify exoplanets according to their Jupiter-like or Earth-like characteristics. There are massive Jovian-type planets referred to as Jupiters and less massive Jovian planets called Neptunes. There are also planets whose masses are up to 10 times that of Earth, and these are referred to as super-Earths. Exoplanets also differ depending on how far they are from their parent sun. Planets that orbit close to their sun are called hot planets and have extremely high temperatures, while planets farther away are called cold planets due to their colder temperatures.

21.5 The Circumstellar Habitable Zone

Today we know that planetary systems are common in the universe. In order for an exoplanet to support life as we know it, the planet would have to be found in a particular area of its solar system called the Circumstellar Habitable Zone. In this region, an Earth-like planet would be at the right distance from its star to be neither too hot nor too cold to be able to maintain liquid water.

Any given star is surrounded by a Circumstellar Habitable Zone. For small and cooler stars the habitable zone is close to the star, and for larger and hotter stars the habitable zone is farther away.

The Kepler Space Telescope has begun to find terrestrial exoplanets that are about the size of Earth and are in habitable zones, but it is not known if any of these planets have conditions that would allow life as we know it to occur.

The search for life on other planets is an exciting area of astronomy to explore. As technology advances, more and more exoplanets will be identified, and we will begin to discover details about their composition, atmosphere, existence of liquid water, and potential for supporting life.

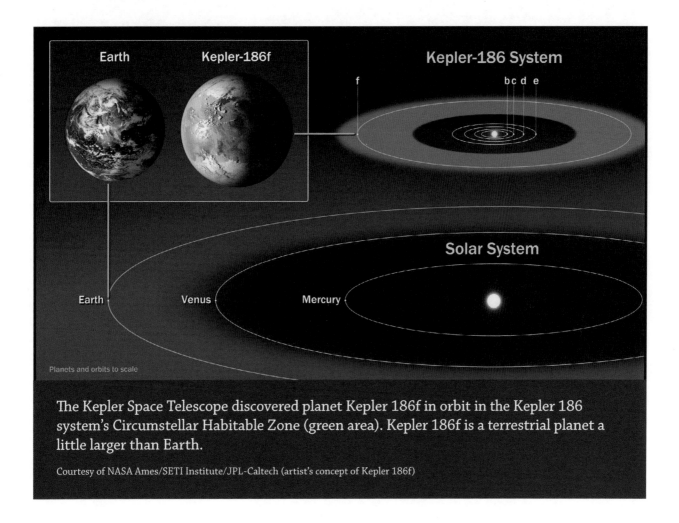

The Kepler Space Telescope discovered planet Kepler 186f in orbit in the Kepler 186 system's Circumstellar Habitable Zone (green area). Kepler 186f is a terrestrial planet a little larger than Earth.

Courtesy of NASA Ames/SETI Institute/JPL-Caltech (artist's concept of Kepler 186f)

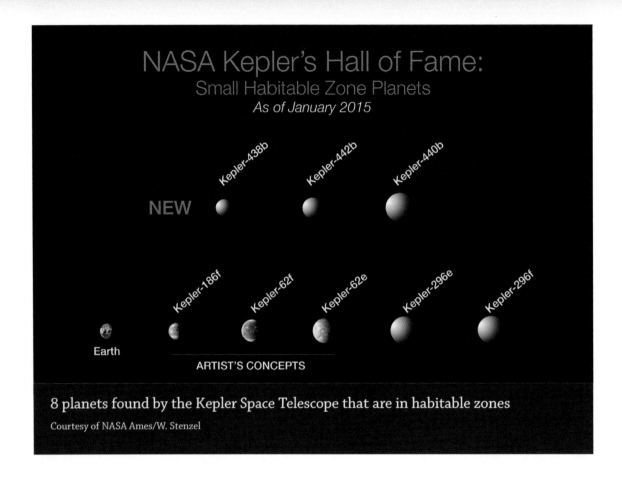

8 planets found by the Kepler Space Telescope that are in habitable zones
Courtesy of NASA Ames/W. Stenzel

21.6 Summary

- Proxima Centauri in the Alpha Centauri system is our nearest stellar neighbor.

- The distances of stars are measured in parsecs. One parsec equals 206,260 AUs.

- The stars that appear brightest and largest from Earth are not the closest stars.

- Extrasolar planets, also called exoplanets, are planets that orbit stars outside our solar system. Many stars are confirmed as having exoplanets.

- In the area called the Circumstellar Habitable Zone, an Earth-like planet would be at the right distance from its star for the presence of liquid water to be possible.

Chapter 22 Working Together

22.1 Introduction	224
22.2 Collaborating	226
22.3 Global Collaboration	228
22.4 Summary	232

Conclusion

Photo courtesy of Gemini Observatory/AURA

22.1 Introduction

In this book we've explored a variety of tools scientists use to study chemistry, biology, physics, geology, and astronomy. We saw how the typical chemistry lab has glassware, balances, safety equipment, and specialized instruments that measure mass and chemical bonds.

We also saw that the biology lab has some of the same equipment as the chemistry lab and how biologists work both inside a lab and outside in the field or in the ocean.

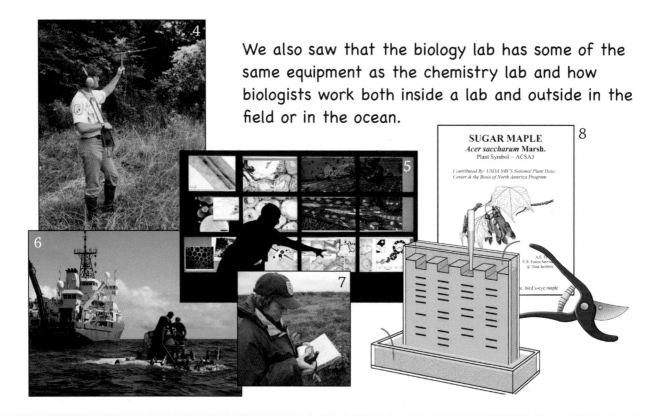

Conclusion—Chapter 22: Working Together

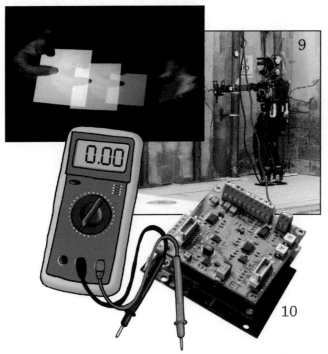

We saw how the physics lab has tools for measuring force, speed, and time and how mathematics and modern computers are vital for collecting and analyzing data.

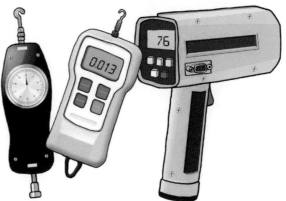

We also saw how geologists use both basic and advanced tools to examine rocks, measure volcanoes, and observe changing landscapes with satellites.

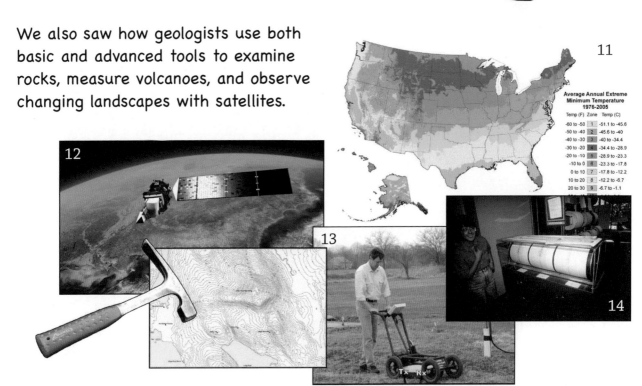

Credits: **Previous page** — 1. Jim Yost/NREL; 2. Warren Gretz/NREL; 3. DOE; 4. John & Karen Hollingworth/USFWS; 5. Dennis Schroeder/NREL; 6. Gulf of Alaska 2004/NOAA Office of Ocean Exploration; 7. Steve Hillebrand/USFWS; 8. USDA/NRCS National Plant Data Center & the Biota of North American Program
This page — 9. DARPA; 10. CSIRO; 11. USDA; 12. NASA's Goddard Space Flight Center; 13. USDA/ARS; 14. USGS/by J. K. Nakata

And finally, we saw how astronomers use telescopes, satellites, space probes, orbiters, rovers, and other spacecraft to explore the cosmos.

Scientific investigation has come a long way from the investigations of early scholars. Today, scientists have access to more tools and more technology than ever before, and this means that scientists can study many areas of the globe and the cosmos for the first time. Scientists can now see into deep space, explore ocean depths, observe atoms, and measure movements on the nanometer scale. Not only can individual scientists use tools and technology to help them make new discoveries, but today scientists can also use the internet to work with other scientists in other universities, cities, and countries. The ability to work together and collaborate means that scientific discoveries can happen at a faster and faster pace as new information and new ideas are shared quickly between labs and as great quantities of data are gathered and analyzed by computers.

22.2 Collaborating

Before the age of the internet, the ability to work with other scientists was limited to phone calls, mail delivery, and travel by car, boat, train, or plane. Scientists would find out about new and exciting projects or ideas from other scientists when they attended a scientific conference or opened up their favorite scientific journal. Scientific projects were historically carried out by an individual scientist or a small group of scientists in a single physical location.

Credits: 1. NASA/JPL-Caltech/Malin Space Science Systems; 2, 3, 5. NASA; 4. NASA/JPL-Caltech

In the early 1800s Gregor Mendel worked alone growing and counting peas because he was curious about why some peas are smooth and some wrinkled. As a solo investigator his work laid the foundation for the field of genetic research. Few people know that for Mendel his work was his life.

Today, scientists can collaborate instantaneously via email or on websites, and this has had a tremendous impact on how scientists share information and ideas. Scientists still attend conferences and travel to physical locations to swap information and work in the labs of their collaborators, but with the internet they can now snap a photo of an erupting volcano and send it immediately to a colleague or upload a set of data from Antarctica to their lab in Italy. Scientists from different countries and from different scientific disciplines can watch in real time as satellites gather data from deep space or as GPS devices track the movement of sharks in the ocean.

Not only can scientists collaborate in real time with other scientists across continents, but solutions to scientific problems can be solved through scientific crowdsourcing, also called crowd science or citizen science. Crowdsourcing is a way to use the internet to enroll thousands of non-professionals and amateur scientists from all over the world who want to contribute money, ideas, or support. Today, crowdsourcing is being used in scientific areas for research such as solving mathematical problems, processing radio signals, and calculating protein structures. Crowdsourcing is also used in non-science areas to raise money and help for business projects.

One way scientists have begun to solve problems through crowdsourcing is to create computer games for science. For example, gamers who are good at solving problems in video games can use their talents to help cure diseases. In 2011 gamers were able to figure out the structure of an important protein involved in AIDS. The players used an online computer game called Foldit to solve the structure of a protein called M-PMV retroviral protease. Scientists had been struggling for years to figure out what this protein looked like, but it only took the gamers three weeks to solve the problem!

Playing AstroDrone
Courtesy of European Space Agency (ESA)

The European Space Agency (ESA) is also using gamers to help with research. ESA has an app called AstroDrone that collects data as a gamer attempts to have a drone connect with a target area as if it were docking with the International Space Station. The second level of the game involves a simulation of navigating the Rosetta orbiter and Philae lander on a mission to a comet. The data collected from people playing these games will be used by robots to learn how to navigate in different environments.

22.3 Global Collaboration

How do you solve big problems like discovering the smallest atomic particles or the DNA sequence of the human genome? One way to do it is to get as many different scientists involved in a project as possible and to have a group of countries fund it.

We saw in Chapter 10 how a group of countries have formed CERN to create the Large Hadron Collider for studying the smallest parts of an atom. This collaborative project was launched in 1954 and originally included 12 European countries. Today CERN has over 20 member countries including non-European countries such as Israel and Pakistan.

But CERN isn't the only collaborative project created by scientists from different countries. The Human Genome Project was launched in the early 1990s and began as a partnership between the US National Institutes of Health (NIH), the US Department of Energy (DOE), and several international partners including Japan, China, Germany, and France. It went on to became the world's largest collaborative project in biology as scientists from many countries worked together to discover more about what makes up human DNA.

Sequencing DNA
Courtesy of CSIRO

The human genome is a very long strand of DNA that contains all the information a human body needs to build and maintain itself. A genome is made up of chemical building blocks called bases that occur in pairs called base pairs, with the human genome containing about 3 billion base pairs. The Human Genome Project was begun because scientists wanted to find out the sequence, or order, in which the base pairs occur in the human genome. The sequence of base pairs is directly related to what a human body is like and how it functions. Because of the size and complexity of a strand of

DNA illustration showing base pairs
Courtesy of Office of Biological and Environmental Research of US Department of Energy Office of Science

DNA, it would be difficult for any individual scientist or lab to sequence it, but because many scientists worked together, the human genome was successfully sequenced in 2003. Today the project has helped researchers discover more about genes, which are groups of base pairs that carry hereditary information from parents to offspring and coordinate the functions of tissues and organs. Scientists have discovered over a thousand genes that can cause disease and are working on ways to use this information to create new therapies. The full sequence of the human genome is now available online for both researchers and the general public.

Another area of global collaborative research is space research. Because space research is so costly and involves such complicated technology, countries often operate group projects. The European Space Agency (ESA) is made up of 20 countries that share their scientific knowledge and financial resources to explore space, study Earth from space, and provide satellite-based services, among other projects. ESA also joins with other agencies from around the world, such as NASA and JAXA (Japan Aerospace Exploration Agency), in different missions and projects. Sharing resources in this way has helped greatly with the rapid advancement of space exploration.

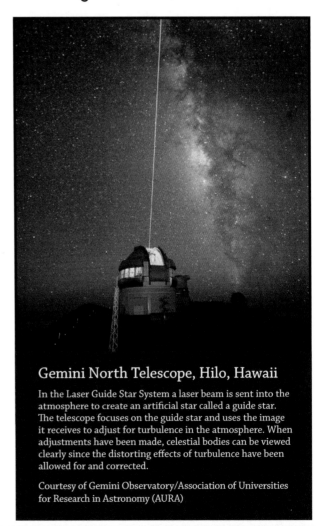

Gemini North Telescope, Hilo, Hawaii

In the Laser Guide Star System a laser beam is sent into the atmosphere to create an artificial star called a guide star. The telescope focuses on the guide star and uses the image it receives to adjust for turbulence in the atmosphere. When adjustments have been made, celestial bodies can be viewed clearly since the distorting effects of turbulence have been allowed for and corrected.

Courtesy of Gemini Observatory/Association of Universities for Research in Astronomy (AURA)

With a single telescope you can view millions of stars. But to observe as many stars as possible in a massive universe is an overwhelming task for any single astronomer. To map, count, and observe the many stars and other celestial bodies in the cosmos, it's better to work together, and this is what several countries have done in the Gemini Project. The Gemini Observatory is an astronomical observatory built by several countries including the United States, Canada, Australia, Argentina, Brazil, and Chile. The Gemini Observatory has two telescopes, one located in Chile in the Chilean Andes and the other located on Mauna Kea, a dormant volcano on the island of Hawaii. Using these two telescopes, astronomers are able to observe almost all of the night sky.

Two galaxies imaged by the Gemini South Telescope, Cerro Pachón, Chile
Courtesy of Gemini Observatory/AURA/Travis Rector, University of Alaska, Anchorage

The International Space Station (ISS) is another example of a collaborative project between countries interested in exploring space. The ISS was created as a multi-country and multi-cultural program between the United States, Russia, Canada, Japan, and the countries in the European Space Agency who wanted to develop a research laboratory in space. The ISS is operated by an international crew and serves an

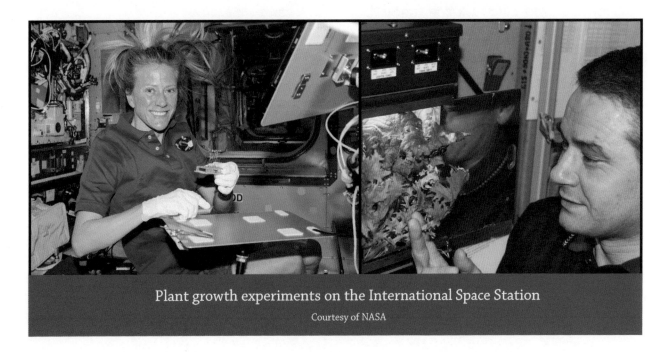

Plant growth experiments on the International Space Station
Courtesy of NASA

international scientific research community. Scientists from all over the world can use and analyze data collected from the ISS. Some scientists even take a trip into space to conduct research on the space station itself.

Advances in technology have made it possible for scientists to visualize everything from extremely tiny to extremely vast aspects of our planet and the universe. Scientists are now able to collect, process, and analyze unimaginable amounts of data. The combination of technological advances, including the internet, and collaboration among scientists and citizens is leading to new discoveries. As each new discovery is made, it lays a foundation for many more discoveries. You may be a collaborator in the next great discovery that makes life better for humanity or helps us reach the stars!

22.4 Summary

- The tools of science are shared between disciplines.
- Modern scientists can collaborate by email and the internet.
- To help find solutions to tough problems, scientists use ideas like crowdsourcing and computer game playing to have many people join together to create large amounts of data for a project.
- Globally, scientists collaborate on big projects such as sequencing DNA, mapping the stars, and exploring space.

Appendix: Math Solutions

Chemistry Math Solutions

You Do It! Ch. 4, p. 40

We know that 84 grams of baking soda equals one mole (page 38).

We also know that one mole of baking soda neutralizes 1 mole of acetic acid (page 33). Looking at the illustration on page 32, we can see that there are the same kind and number of atoms in the molecules before the reaction and in the new molecules that result from the reaction. We can write the reaction as:

$$CH_3COOH + NaCO_3H \longrightarrow NaCH_3COO + CO_2 + H_2O$$
acetic acid + baking soda　　sodium acetate + carbon dioxide + water

Because we know how the atoms and molecules behave during the reaction, we can see that 2 moles of baking soda will neutralize 2 moles of acetic acid, 3 moles of baking soda will neutralize 3 moles of acetic acid, 0.5 moles of baking soda will neutralize 0.5 moles of acetic acid, and so on.

The questions in this *You Do It* section ask how many moles of acid are neutralized by 84 grams, 42 grams, and 168 grams of baking soda. In each case all we need to do to solve the problem is to convert grams of baking soda to moles of baking soda. Since an equal concentration of baking soda will neutralize an equal concentration of acetic acid, we know that the number of moles of baking soda and the number of moles of acetic acid will be the same.

To convert grams to moles we use a "conversion factor." A conversion factor states mathematically the relationship between two quantities. For baking soda we can write our conversion factor as:

$$\frac{1 \text{ mole}}{84 \text{ grams}} \quad \text{or} \quad \frac{84 \text{ grams}}{1 \text{ mole}}$$

(Continued on next page.)

Chemistry Math Solutions (cont.)

To solve these problems, multiply the number of grams by the conversion factor.

1. If it takes 84 grams of baking soda to neutralize a beaker of acetic acid, how many moles of acetic acid do you have?

$$84 \text{ grams} \times \frac{1 \text{ mole}}{84 \text{ grams}} = 1 \text{ mole}$$

There is 1 mole of acetic acid that is neutralized by 84 grams of baking soda.

2. If if takes 42 grams of baking soda to neutralize a beaker of acetic acid, how many moles of acetic acid do you have?

$$42 \text{ grams} \times \frac{1 \text{ mole}}{84 \text{ grams}} = 0.5 \text{ mole}$$

0.5 mole of acetic acid is neutralized by 42 grams of baking soda.

3. If if takes 168 grams of baking soda to neutralize a beaker of acetic acid, how many moles of acetic acid do you have?

$$168 \text{ grams} \times \frac{1 \text{ mole}}{84 \text{ grams}} = 2 \text{ moles}$$

2 moles of acetic acid are neutralized by 168 grams of baking soda.

Appendix: Math Solutions (cont.)

Physics Math Solutions

Problem 1   Ch. 12, p. 130

1. $\dfrac{10 \text{ m/sec.} - 25 \text{ m/sec.}}{|5 \text{ seconds} - 0 \text{ seconds}|} = \dfrac{-15 \text{ m/sec.}}{5 \text{ seconds}} = -3 \text{ m/second}^2$

2. $\dfrac{40 \text{ m/sec.} - 10 \text{ m/sec.}}{|10 \text{ seconds} - 0 \text{ seconds}|} = \dfrac{30 \text{ m/sec.}}{10 \text{ seconds}} = 3 \text{ m/second}^2$

Problem 2   Ch. 12, p. 131

v_i = 20 km per hour
v_f = 6 km per hour
t_i = 0
t_f = 0.10 hour (6 minutes)

What is the acceleration?

$a = \dfrac{v_f - v_i}{|t_f - t_i|} = \dfrac{6 \text{ km/hr} - 20 \text{ km/hr}}{|0.10 \text{ hr} - 0 \text{ hr.}|} = \dfrac{-14 \text{ km/hr}}{0.10 \text{ hr}} = -140 \text{ km/hr}^2$

Problem 3   Ch. 13, p. 137

No. The center point doesn't travel any distance.

Problem 4   Ch. 13, p. 139

1. Farther out
2. Both positions have the same rotational speed
3. It doesn't move. It has no speed.

More REAL SCIENCE-4-KIDS Books
by Rebecca W. Keller, PhD

Focus Series unit study program — each title has a Student Textbook with accompanying Laboratory Workbook, Teacher's Manual, Study Folder, Quizzes, and Recorded Lectures

Focus On Elementary Chemistry
Focus On Elementary Biology
Focus On Elementary Physics
Focus On Elementary Geology
Focus On Elementary Astronomy

Focus On Middle School Chemistry
Focus On Middle School Biology
Focus On Middle School Physics
Focus On Middle School Geology
Focus On Middle School Astronomy

Focus On High School Chemistry

Building Blocks Series yearlong study program — each Student Textbook has accompanying Laboratory Notebook, Teacher's Manual, Lesson Plan, and Quizzes

Exploring the Building Blocks of Science Book K (Activity Book)
Exploring the Building Blocks of Science Book 1
Exploring the Building Blocks of Science Book 2
Exploring the Building Blocks of Science Book 3
Exploring the Building Blocks of Science Book 4
Exploring the Building Blocks of Science Book 5
Exploring the Building Blocks of Science Book 6
Exploring the Building Blocks of Science Book 7
Exploring the Building Blocks of Science Book 8

Super Simple Science Experiments Series

21 Super Simple Chemistry Experiments
21 Super Simple Biology Experiments
21 Super Simple Physics Experiments
21 Super Simple Geology Experiments
21 Super Simple Astronomy Experiments
101 Super Simple Science Experiments

Kogs-4-Kids Series interdisciplinary workbooks that connect science to other areas of study

Physics Connects to Language
Biology Connects to Language
Chemistry Connects to Language
Geology Connects to Language
Astronomy Connects to Language

Note: A few titles may still be in production.

Gravitas Publications Inc.
www.realscience4kids.com

Made in the USA
Columbia, SC
14 April 2018